LASSETER'S DREAM OF MILLIONS

NEW LIGHT ON
THE LOST GOLD REEF

FRED BLAKELEY

Western Australia Border
Ehrenberg Range
Mt Liebig
Haast Bluff
Lake MacDonald
Macdonnell Ranges
Alice Springs
Petermann Range

LASSETER'S DREAM OF MILLIONS

NEW LIGHT ON
THE LOST GOLD REEF

FRED BLAKELEY

Edited by
FRANCES WHEELHOUSE
and
MARY MANSFIELD

TRANSPAREON
PRESS

DREAM MILLIONS

First published in 1972 by
Angus and Robertson (Publishers) Pty. Ltd.,
102 Glover Street, Cremorne, Sydney.

LASSETER'S DREAM OF MILLIONS

Transpareon Press
P.O. Box 126, Hornsby, N.S.W., Sydney, Australia 2077

First published new edition 1984

Cover Designed by Judy Hungerford

National Library of Australia
Cataloguing-in-Publication Data

Blakeley, Fred, 1882-1962
Lasseter's Dream of Millions

Includes Index

ISBN 0 908021 05 4

1. Lasseter, Harold Bell, d. 1931
2. Prospecting—Australia, Central—History
3. Gold mines and mining—Australia, Central

1. Wheelhouse, Frances, 1924-
11. Mansfield, Mary
111. Title

919.42'0442

Printed by The Dominion Press-Hedges & Bell,
Maryborough, 3465

Contents

Foreword

THIS BOOK, which I prepared for publication under the author's direction from the original manuscript, is an account of the Lasseter Expedition of 1930 (from which Lasseter did not return), as set forth by its leader, Fred Blakeley.

I first met the author early in 1940, and was very much struck with his story of the search for Lasseter's Reef—which he claimed did not exist—and of the many people who had risked their lives and spent their money looking for the fabulous gold.

What was even more interesting was Blakeley's firm and unshakeable conviction that Lasseter did not die in the centre of Australia as generally supposed. That—to me—was quite a shock. Like many others I knew only the universally accepted story that he had perished in the desert. It seemed incredible that Blakeley's theory could be true. However, after some discussion, it was arranged that I should read the manuscript with a view to editing it for publication; and I found that the story unwound in a straightforward, sensible way.

We worked together somewhat spasmodically till eventually the actual narrative was finished. There was to be a final chapter which Blakeley called "The Summings Up"; he meant to write it but never did. This was to show certain reasons why Lasseter could not have made the trip on which he claimed to have found the wonder reef.

I lost touch with Blakeley after the main manuscript was finished, when he was to gather the material for the "last chapter". Time passed, and I was absorbed in other things and I did not know that he had died in 1962 till long after the event. Now I am passing on his story. Some minor corrections of style have been made to the manuscript and some extraneous and discursive material that was impeding the flow of the narrative has been deleted; but substantially I am presenting this account just as Blakeley wrote it.

I have been most ably assisted in the final preparation of the manuscript by Miss Frances Wheelhouse.

M.M.

1

Ten Miles of Gold

ONE DAY while we were having lunch together, my city friend Malcolm Stanley spoke of yet another expedition that had just started off in search of Lasseter's reef of gold. Stanley said, "Why don't you write an account of the expedition you led in 1930? Tell the story just as you have related it to me from time to time."

At first I refused, saying that I did not wish to injure Lasseter's wife and four children. Nor did I want to enter into competition with two books that professed to tell the story of the expedition and Lasseter's end.

Stanley persisted, saying, "I think it would be a great mistake to let the deception go on. This mythical reef has already cost hundreds of thousands of pounds, sickening disappointment, and the loss of lives. It is not 'past history'. It is up to you to tell the story as you believe it to be correct."

Here, then, is the story. . . .

I had received a letter from my brother, Arthur Blakeley, who was Minister for Home Affairs in the Scullin Government. John Bailey, at that time president of the Australian Workers' Union, had written to Arthur asking where I might be found. I happened to be in Sydney so I went to see Bailey who told me that a chap had drifted into his office with a very likely story of a gold reef he had found in Central Australia. I agreed to test the chap out. The following Saturday Bailey introduced me to Harry Lasseter. We always called him Harry or "Possum" but in the records of the expedition he signed himself as Lance Harold Bell Lasseter.

Some plain talking was done in Bailey's office and then Lasseter and I went into Hyde Park, where I heard his story. Knowing that the man was after financial support I made notes, and after two hours I agreed to meet him in the same place on the following Saturday. I spent four Saturday afternoons with him; then I went to Bailey and asked him to be in his office the following Saturday. There Lasseter and I met him and I made my report. I told Bailey that I had taken a month to consider Lasseter's story. There were a lot of gaps in it but I felt that there might be something in the yarn. Lasseter had explained the gaps by saying in reply to my questions, "If I told you that, you'd know as much as I do, and you wouldn't want me."

These conversations had left me suspicious of Lasseter's story and of his knowledge of mining and bushcraft. I told Bailey of my doubts but said that I was ready to take out an expedition to my old stamping ground. To test out the fellow's claims would cost about £300. Bailey said, "I can get that amount in this office", and he and I put in £50 each. On the Monday evening six of us met in Bailey's office and we formed a syndicate. We went into details of how the expedition was to be equipped and adjourned the meeting until the following Monday night. In the meantime the news had got around and several of our friends were clamouring to be admitted to the syndicate. We six had an interim meeting and agreed to let our friends attend the next meeting, but to our surprise over twenty persons turned up. The first business of the meeting was to hear Lasseter tell his story. I knew it pretty well by this time.

"When I was about seventeen years of age," ran Lasseter's yarn, "I heard of a new ruby field that had been found in the Macdonnell Range and that hundreds of men were rushing there. I was in North Queensland and decided to join in the rush. I considered myself to be a good enough bushman to cut across Cloncurry and Paul's [*sic*] Creek, then follow the telegraph line to the Macdonnell Range, considerably over a thousand miles."

As Lasseter would have been about forty-seven in 1930 this would have been thirty years before. The distance from Cloncurry to the Macdonnell Range is nearer six hundred miles, but even so I was doubtful whether a boy of seventeen could have managed it with only one horse over the country he would have to travel.

"The journey," said Lasseter, "took me a lot longer than I had bargained for. When I arrived at the 'ruby' field I found that the boom had burst and practically all the miners had left. The rubies had turned out to be garnets. I called in at Alice Springs telegraph station and got supplies, for I intended to go right through to Carnarvon in Western Australia. I struck a western line by compass, and it was while on this part of the journey that I came across the reef. I did not have any knowledge of mining then, but I could not help looking at the reef, it ran in the same direction as I was going. I could see it outcropping here and there for a couple of miles. I noticed stones in a heap just as if someone had pulled them from the reef and thrown them together. I examined the stones and could see long flakes of gold and lots of small specks in the stone. I did not take any of this but followed the line of lode to where the reef outcropped about five feet above the ground. I again examined the stone and could see gold all through it. I gathered a small bag of about three or four pounds in weight. I continued to examine the reef as I rode along. It was about ten miles long to where it disappeared under the surface, with an average width of eight to twelve feet. Everywhere I examined it I saw gold."

Lasseter then told how he was undecided whether to go back or go on. I questioned him and he admitted that it would have been impossible for a man with only one horse to have made the 1,200 mile journey to Carnarvon. He said that he had very little idea of distance but he knew that long before he got back to Alice Springs he would be out of tucker, so he decided to force his way on. Yet the further he went the worse grew the country. It was now all sandhills and salt lakes.

"A few days after passing the reef country," Lasseter went on, "my horse died. And I was in a bad state myself. After wandering for I don't know how many days I was picked up by a surveying party. All I had when they found me was part of the bag of specimens. After some weeks the surveyor who found me, Mr Harding, brought me into Carnarvon. At this time the Western Australian goldfields were booming."

Lasseter said that he joined in the rush to get practical experience but kept in touch with Harding. Three years later he and Harding got a plant together and after many months relocated the reef. They took samples from practically the whole length, but their supplies ran out and water was hard to find. Harding took the position of the reef with his sextant and he and Lasseter returned to Carnarvon, some six to seven hundred miles from the reef.

"We could not get anyone interested in our story," Lasseter told the gathering. "Our specimens assayed three ounces to the ton, but this was not very attractive in those boom days. Mr Harding hawked our show to Adelaide and Melbourne but could not find anyone to give financial support. We drifted apart and then Harding died. I am sure I can locate the reef again. This time I propose to go out on my original track, starting from Alice Springs."

At this meeting Lasseter said very definitely that his reef country was *about three hundred miles* from Alice Springs, but that from Carnarvon it would be over seven hundred miles.

Bailey then told the meeting how I had examined Lasseter. The details, said Bailey, did not really matter as we would be working on a position taken by a competent surveyor. He explained that I had taken five or six weeks to think over the story, but I was prepared to put in £50 towards the £300 needed.

Someone asked whether it would be in order to ask questions. Bailey replied that, as it was an informal discussion, anyone could talk or ask questions. Although no resolution was put to the meeting, it was practically settled that as £300 did not seem enough capital, a company should be formed with £1,000 capital made up of £20 shares, no person to have more than one share.

From then on it was hard to keep track of events in Sydney. From our syndicate with its £300 had come the suggested company with £1,000 invested; but with still more clamouring for shares, after two or three meetings a company was formed, Central Australian Gold Exploration Coy Ltd, with a capital of £5,000. Lawyers were introduced and there were many legal arguments. Then two very live wires came in to help, Errol Coote and Mr Charles Lexius-Burlington, and things fairly hummed. All sorts of schemes were suggested, but the clause of one share to each person still held, though people were applying for shares for members of their families and their relations. I lost track of this side of the company's affairs.

As a result of Burlington's efforts the manager of Thornycroft Motors announced that his company would lend a six-wheel, two-and-a-half-ton truck for six months. Then an agent of the Atlantic Oil Company gave six hundred gallons of petrol and a hundred gallons of oil to run the truck. The Central Australian Gold Exploration Company decided to buy an up-to-date light flying machine.

Everyone was riding on a wave of enthusiasm, for here was something the country badly needed. Harry Lasseter, the little nuggety man, seemed to have shown a way to break the Depression. The patriotic spirit was simply grand, and it was a great pity things turned out the way they did.

The organization of the expedition took about six weeks, and during this time Lasseter was paid by the company £5 a week waiting time. Just as everything was ready for the start Lasseter took objection to having Burlington as a member of the expedition, saying that he would not go if the company insisted on Burlington being included. His objections were noted and Burlington was replaced by Philip Taylor, an Englishman from de Havilland's works in Melbourne. Taylor had his long service leave due and, as he was contemplating marriage with an Australian girl, he could see the advantage of a nice cheque to top up his holidays and honeymoon. Another man introduced to the company at this time was Captain Blakiston-Houston, aide-de-camp to Lord Stonehaven, the Governor-General. Blakiston-Houston was on holidays at the time and was anxious to go with the expedition for two or three weeks' Central Australian experience.

The party, then, finally settled upon by the company consisted of myself as leader, Harry Lasseter as guide, George Sutherland, prospector and miner, Philip Taylor, engineer and driver, Errol Coote as pilot of the plane, and Captain Blakiston-Houston who was given the title of explorer.

2

The Expedition Sets Out

THE PARTY left in dribs and drabs, but everything had been carefully planned. The Thornycroft truck, fully loaded, was placed on a railway truck at Darling Harbour. It had two tons of provisions mostly provided by McIlwraith's grocery stores at cost price, a full kit of blacksmith's tools, a boring plant, a condensing plant and an oxy-welding plant, all secured at cost price. The New South Wales Government gave free transport as far as Broken Hill for the truck and party. The Silverton company granted a similar concession but the South Australian Government was not interested; so Lasseter, who accompanied the truck through to Alice Springs, hired a driver to Quorn, the head of the North-South line owned by the Federal Government, which also gave free transport to the expedition. George Sutherland and I went by passenger train and caught up with Lasseter at Quorn.

We were lucky enough to catch a special cattle train from Quorn to Alice Springs, and so saved two days on the schedule. This enabled us to have the sorting of the load well in hand by the time the others arrived. But let me give you a tip: Don't ride on a cattle train if you like a comfortable journey, for you will never forget going over the sandhills! The railway is laid right on the surface of the sand and there are no cuttings, just some sixty ridges to cross and most of these are quite steep and fairly close together.

Cattle trucks are not coupled together like passenger cars and bump terribly when one is riding the tail of the train, as we were, in a self-contained sleeping and cooking car. There was no sleep that night, everyone felt as though they had been rolled down a stony hill in a barrel. I think the train crew had put on something special for our benefit for, when they came off shift, they were laughing and doubtless thought they had had a great joke with three "new-chum" travellers.

The whole population of Alice Springs knew we were coming, but not one white resident turned out to see what the new expedition looked like. They were used to such expeditions. I think we were the twenty-second or twenty-third to arrive with full organization and plant on a similar quest for treasure.

When the other members of our party arrived some four days later, two by plane, one by passenger train, did they come into our camp and start to work as we had? Not on your life. They went to the hotel, dressed up in their pretties and tried the looking-on stunt. But the second day that was changed and they were all working and sorting out stuff; three or four tons of it of which we could only carry half. It was a big job, for we intended to make a depot three hundred miles away. We were taking with us one ton of provisions, camp equipment, oxy-welding, condensing plants and the two-way wireless plant that weighed over one hundredweight. We also had blacksmith's and other tools that weighed just about a ton.

We intended to take on about a ton of petrol and oil, and later to pick up at least half a ton of water, so the work of sorting out things took about eleven days. Here is where the Sydney office showed their ignorance of what a huge job it was to get the six-wheeled truck packed so that it had any chance of getting through. This truck, on city roads, would do from fifteen to seventeen miles per gallon; on our track we averaged about three miles to the gallon. Stupid wires from headquarters came every day asking for explanations of what was delaying us. These annoyed me very much as they all had to be replied to, and this burning of the telegraph wires was a costly matter.

I am doubtful if we could have been ready a day sooner in the circumstances. Everyone, including the hired hands, was glad when knock-off time came. I knew the job I was taking on, and knew it would be no use remembering something later on! It had all to be thought out first, and, because of this, I had Captain Houston double-check everything. Nothing was left to chance, for I knew the country we were bound for.

It was at this stage that I hired Fred Colson and his light motor truck at £3 per day, providing him with petrol and tucker. Over this action I came in for some undeserved criticism.

I had done some deep thinking about driving this load of five or six tons out to our depot. It would take at least three weeks for the big truck to reach Ilbilba, three hundred miles west of Alice Springs. If we took enough provisions we would not be able to carry enough petrol for the return journey. Besides, I was frightened of the short amount of time we had; already we were starting out fully six weeks too late, and we would probably have the hot weather on us before we got a start. Motor trucks are not constructed to work in such heat and loose sand, no matter what their make, for when the ground temperature reaches 175 degrees it is almost impossible to keep water in the radiator. The Sydney crowd could not understand these things, not even when I wrote or sent lettergrams explaining why I had to have the light auxiliary truck. One bumptious member of the company mischievously sowed the seeds of discontent, though I did not know it at the time, by insinuating that I was going to double-cross the others and work in with Colson when Lasseter found the reef.

It would not have been so bad if he had just exercised his flights of fancy on the Sydney crowd, but he went further and aroused the misgivings of the naturally suspicious-minded Lasseter. Later, when I found this out, it was a wonder that I did not commit murder!

I was no stranger to Alice Springs. I knew many people there and had some fine old friends of over twenty years' standing. One of these, and the most outstanding man in Central Australia, was Mr Allchurch, postmaster and magistrate, who could always be approached. He was the friend of everyone and especially of the natives who always got a fair deal from him. He spent practically the whole of his working life at this station. I made arrangements for him to meet Lasseter and, when I took him to the telegraph station, Lasseter made his first noticeable mistake.

I introduced him to Mr Allchurch and others; then Harry Lasseter began talking. "Yes," he said, "I remember those old buildings quite well!" The buildings he referred to were partly stables, harness and general store-rooms. Harry said, "I came just after the great ruby rush was over and got supplies."

Mr Allchurch said, "That is a good bit over twenty-five years ago, mate."

"Yes, I know that, I came here from Queensland to mine for rubies."

"Well," said Mr Allchurch, "you don't remember those buildings for they are only a little over twenty years old."

But that was not warning enough and Lasseter went on to tell how he had come there and got fresh supplies.

Mr Allchurch said, "From this station?"

Harry said, "Yes."

"Well, old man, all I can say is that you dreamt it, for I have been station-master here for forty years and I am sure that never once did I issue stores to anyone. And I think if you stayed here I would remember you, because you must have been very young."

I stopped the cross-examination by taking Harry away, and got into him for talking such rot. Harry did not know that the big store in Alice Springs was there in those days and maybe a larger concern than it is today. He was not even aware that the telegraph station and the town were two miles apart. One evening Mr Allchurch was very scathing and told me I was a fool to believe anything the fellow said.

After that I started to pay more attention to what Harry *did* say, but he was cautious now and held his tongue. After a day or two I gave him the benefit of the doubt, for anyone could make mistakes with thirty-odd years separating his visits. Even between my own visits, I hardly knew the place.

The criticism of the big truck was funny to hear. Almost everyone predicted that it would not get a hundred miles. They said that it was too big and heavy, and too low-set, but they did not know what a six-wheeled Thornycroft could do. They lived to learn differently: only camels could go where the old Thornycroft couldn't.

Before we left Alice Springs I had to go before the Resident Magistrate and make a declaration concerning all the firearms we were carrying. Several busy-bodies in the cities had raised a hue and cry of how the expedition was like a travelling fort and that we were all armed to the teeth and they heard that we were going to slaughter all the natives.

Then a member of the Opposition, in the Federal House at Canberra, made an attack on me about the firearms—because I happened to be a brother of the Minister for Home Affairs. I had to wire my brother and give the number of arms carried by the expedition. Each man carried one revolver, there were two sporting rifles and one shotgun. I have always been a staunch protector of our natives, and added in my declaration that I would not barter or collect any native weapons or remove anything from the graves of natives.

The reason why every man carried a revolver was in case he met a mad brumby camel, for any of us might have to be away from the truck for hours. I had one experience with a bull camel in season and, had I not had my revolver handy, I might have got my wings.

On the afternoon of 21st July 1930 we started off from the Alice with the best-equipped turnout that had ever been in this country. We had only gone two hundred yards when we ran into trouble. We were crossing the dry sandy bed of the river and Phil, who had not driven this kind of a vehicle before, found the steering was different so he got across the beaten wheeltracks and, of course, was stuck. However, he soon backed out, for it was a powerful thing in reverse. We ran out forty-five miles that afternoon in less than three hours.

That night at about 8.30 we were all surprised to see the Government Resident, Mr Carrington, drive his car up. His first remark was, "Well, you got further away than we thought possible."

After we left Alice Springs a telegram had come from our Sydney office and as it was an alteration to our wireless code, he had thought it important enough to try to deliver it.

I thought this jolly fine of him and added many thanks but this was Central Australia where such things are quite ordinary, for there is a fellowship out there that does not exist in other parts of Australia.

The irony of Mr Carrington's long drive was that the code was never used, for no one understood the complicated thing that had been palmed off on us as a wireless: I will tell you more about it later.

3

Out into the Centre

NEXT morning we were all astir early, for one night in camp had brought about a better understanding. Everyone was happy and we were at last on our way.

Of the seven members of the party, George Sutherland was a practical goldminer and deputy leader, and a happy fellow. If he was not singing he was telling some yarn or other. The only thing against his yarns was that we seldom heard the end of one because his voice suffered from statics and faded away. He was a tall, lanky, tow-headed Australian Scot, and every time the old truck got a quick move on he would start singing "Off to Bonny Scotland".

He sat on the left hand of the driver, Phil Taylor, a young Englishman, dark-complexioned with a quick, fiery temper. It was funny hearing Phil trying to swear. He could not get the flow that Australians achieve—he was never in rhyme or time—but he was a good fellow, one of the best, and as time mellows the memories of that gruelling experience I doubt if such a driving feat will ever be repeated. He could take that truck into anything and get it out again. Besides, he was a fully qualified air and motor mechanic and a clever blacksmith and oxy-welder. He did some delicate repair work to the motors and planes. One thing that took him a long time to get over was that he could not have his bath when he got covered with sand and grease, as he often did, for although we carried plenty of water we never knew how far away we were from the next supply.

Then, of course, there was Harry Lasseter. He was a little nuggetty fellow, darkish complexion, flat chubby face with no nose-bridge. He was partly bald, and his scalp showed many deep scars. He was a man of jumbled moods, and never quite happy; if anything went a bit wrong he sulked. His special pastime was singing hymns, all set to the one tune, for he was a Mormon by faith. He joined that crowd in America and was a naturalized American, but under a different name. Lasseter sat on Sutherland's left side.

Errol Coote, the air pilot, had control of a Gypsy Moth called the *Golden Quest*. He was a man of slight build and dark complexion, who packed short bursts of violent temper. His rigout was laughable. He wore boots like girls wear to ride in at Shows, and had he been compelled to walk ten miles he would have had only the uppers left. He did not possess the right temperament for such a trip since he was jerky and jumpy and a journalist with a vivid imagination who was always on the job. If he saw a dingo he would write of the "snarling and snapping of its teeth". He often told me of the things he could do with the *Golden Quest*, but that machine only gave me a lot of worry and trouble and at one time he delayed the whole expedition for three weeks. He drove Fred Colson's motor car. Fred had lent me this car which also accompanied us so that Captain Houston could return when his time was up.

Captain Blakiston-Houston, as I have said, was the Governor-General's aide-de-camp. He wanted some real Central Australian experience. The Sydney office granted his extraordinary request because it gave the personnel of the party a hoity-toity touch. I had some misgivings about his inclusion, but, after meeting him, we had a long talk and I felt pleased that he would be with us for the first two or three weeks. I soon recognized that this man was no stranger to roughing it as he had been in charge of a section of an expedition that attempted to conquer Mt Everest. He was an English-johnny type with an Oxford accent, tall and with a fairish complexion and military bearing. He was an English soldier from the crown of his head to the tips of his toes. His knowledge and assistance proved very valuable in the breaking-up and loading and we went on his advice as to how each lot should be placed on the truck. This is where his Everest experience came in. He was never behind when the sand-prickled and thorn-filled matting had to be laid. He carried his share of water and only sat down when everything had been done for the night. Sutherland, Colson, Taylor and myself were really sorry when he had to return.

Fred Colson was a stockily built ball of energy. He had a bricky complexion with a fine, open, pleasant face with rather enlarged nostrils apparently developed from heavy breathing through excessive exertion in the energetic manner in which he did a job. A real man all through, he never took on a job that he could not carry out to the letter. He was as reliable as *Sunrise*, the Chev. auxiliary truck. With him behind I never had any fear that our supplies would not be delivered on time. The lives of those on the big Thorny more or less depended on him. I would even have chanced taking the Thorny further than the safety margin because we all knew that Freddy was behind. He was indeed a great fellow, a typical Central bushman who lived at Alice Springs with his game little wife who was just a notch better than him.

The big Thornycroft broke the track. Freddy came second in his truck, and the car was very often a bad last. When one was standing away from the little column of motors it looked rather imposing and it seemed to us as though the western portion of Central Australia would be conquered this time. Never before had such an efficiently equipped expedition attempted to cross into the unknown mysteries of these elevated, sand-ridged, seared plains.

We were all astir at break of day. It was very cold and every man had his top-coat on and stole a few moments between packing to warm his hands. We were all very happy; snatches of song and whistles were the order of the day. George Sutherland called "Breakfast", for he was the cook that morning and treated us to porridge, then ham and eggs with toasted baker's bread.

At about 7.30 I called "All aboard", and we moved on our western course. There was some semblance of road here, the going was good, and we covered the next thirty-five miles in grand style, arriving at a cattle-station tank about 10.30 a.m. This first eighty miles was the best track on the whole trip. We intended to take on a hundred gallons of water from the tanks. I had asked some "reliable" cow about this water and was told it was good. There were about eighteen inches of thick yellow soup, the tank was nearly empty, and Hamilton Downs cattle station had large numbers of stockmen out mustering the cattle to move them to other water.

I groaned, for we had passed a splendid well twenty-two miles out from the Alice, but I was a bit shy of well water—very often it turned bad after a day or two. We were told that we could get plenty of good fresh water at the tank. We only took on forty gallons which I thought would do for the truck radiators.

But we still had another chance. Archie Giles's station was about twenty-seven miles ahead, slightly off our track, so we left the waterhole of stinking soup; but it took until 3 p.m. to get opposite Archie Giles's homestead.

We had a taste of what sand could be like, and all hands were busy laying the matting. This coconut matting was made especially for this kind of work. It was about eighteen inches wide and fifty feet long and we had four lengths of it. This was Phil's first attempt at guiding the big lumbering truck across matting, and, of course, it ran off a few times. Then we went shovelling out, sometimes jacking up the wheel. If there were any bushes handy we made a bed of them under the wheel.

We made camp about three o'clock in the afternoon, about four miles away from the homestead. We lightened up the auxiliary truck, and Freddy and I went into the station as I had permission from Mr Carrington to pick up a native boy here. I had a covering note to Archie Giles.

When we arrived at the homestead we found it deserted; not even a cupful of water and tanks and wells were both dry. We looked around and found a camel-pad leading over the hills, and followed it on foot for seven miles before we came upon Archie Giles camped at a waterhole. In spite of the trying conditions the old fellow met us with smiles. He was well known to Fred Colson, and I had met him once before. I was pleased that he remembered me; it was twenty-two years since our previous meeting at Anna Creek station.

4

Archie Giles

I WOULD LIKE to describe Archie Giles to you. He is a splendid chap but I find it a delicate matter to speak of him because he is a half-caste, the son of an explorer. Still, I don't think he would mind; he was always proud of his birth.

I think he is the most outstanding half-caste in Australia, respected and loved by all who know him. Never in his life has the colour-line been imposed. Wherever he goes he is always invited to the table for a meal. He has made many trips to Adelaide and has sometimes gone to Melbourne to see the Melbourne Cup race. He could stay at any hotel and not be questioned.

A man of about seventy years of age at this time, he had had a wide experience among the Aborigines of Central Australia. He could speak several tribal languages, and knew the sign-languages, that native writing on the sands, wood or rock. He understood all the hand-signs and smoke-signals; he knew the native religious beliefs and laws.

This portion of his cattle-run is among the rough hills, a part of the Macdonnell Range. Where he was camped were two rock-holes that had a little water, very nearly as bad as the Hamilton Downs tank. It wasn't muddy, but it had a thick, green scum; most of it was polluted by the cattle wallowing in it for over a year, to say nothing of the small birds that bathed three or four times daily.

After our greeting Archie put on the billy and we drank the beef-flavoured tea, which we both thoroughly enjoyed, for we were indeed thirsty after our long walk. I told him of our mission and showed him Mr Carrington's permit. He told me that I was unlucky, as all his smart boys were helping to muster cattle at Hamilton Downs, a neighbour's station. They were moving the cattle about a hundred miles and would not be back for three or four weeks.

So he lent me a man of about forty-five years of age, who had very bad eyesight through the ravages of the dreaded sandy blight. As this was my last chance of getting a native, I accepted him. Mickey knew the country for about eighty miles ahead, but had I known that he was practically blind I would not have taken him. I really wanted a boy who could tell me something about animal tracks and the flight of the birds and insects.

These natives can tell if they are going or coming from water by the water-fly. It is only the sharp eyes of the native that can see these flies. They fly much faster than a bee and are only a quarter the size. The life cycle of these water-flies would be interesting to know if one could get full information. Their mission seems to be flying to the water and filling up; they increase in size considerably when loaded. The natives' eyes are so good that they can pick the loaded fly from the empty one and, step by step, they follow the direction that the loaded ones are coming from until they reach their well. This is smaller than the diameter of a lead pencil. When the natives find a well they look around for more; and, according to the number of wells, they can tell the size of the underground water supply.

I have asked natives what the flies do with their loads of water but could never get one to show me. They just give a couple of hand signs that I could not understand and say, "Him go alonga another hole—sometimes alonga tree—sometimes alonga rock, him all clacked." I asked, "Do you know what they do with the water when they go down the other hole?" They would only laugh at my question and say, "Him get thirsty pretty quick, plenty times all day." I asked, "You show me when him go down other hole." Then with a serious face the native would say, "That way—nothing water," pointing the way the water carriers were flying. "Him water alonga here," pointing round to the many fly wells. They would give some hand or lip sign and point with their lip or tongue in the direction that the flies were going. So that would end the matter, for once you bump against the sign language you know that these sorts of things are not talked about verbally. It evidently has something to do with their spirit land, and it would be bad for the native tribe if a bad spirit heard them talking.

Archie put saddles on four camels and we all rode back to our camp. Fred, of course, picked up his truck. It was just about dark when we got there.

At this early stage of our journey we had fresh meat, baker's bread, eggs, and a couple of bottles intact so I put on our best show for our tea. Old Archie laughed when he got two eggs, and a good thick slice of ham, and a heavily buttered slice of real baker's bread, for in these hard times he lived on damper and meat, such things as butter and jam not being on the ration list.

We had a jolly evening. Archie told us that the country had missed the thunderstorms for three years running and that he had saved a small mob of cows and bullocks, about three or four hundred. The thunderstorms were due in three or four months. If they missed this time, it meant the end of his little mob. I am pleased to say that he got the thunderstorms—eight of them—all worth over one inch of rain. They were happy days for those pioneer cattlemen.

We turned in just before midnight. I felt very happy over the meeting with an old friend; besides, I had been given good directions about the country ahead for about a hundred miles, and the news that about sixty or seventy miles ahead we would find good soakage-water supply. We were up early and breakfasted well, including rolled-oats porridge. It was grand to see our fine old visitor stowing away a good feed. In his tucker bags there were a couple of pounds of butter and some eggs, and best of all a pound or two of tobacco. So we said farewell, getting his blessing and wishes of, "Good luck, leave the girls alone, and don't camp on waterholes and you won't have any trouble with the natives."

From here on we were going into the unknown, as far as wheel traffic was concerned. Archie told us of a spring we could reach the next day, for the water shortage was becoming acute. Mickey, our latest addition, sat on Lasseter's left hand. This made five of the party sitting on the long seat of the big truck and I sat on Phil's right hand, so he had three on his left hand and one on his right. It was a roomy cabin, the seat and luggage carriers overhead made two good berths. Lasseter always slept in the cabin.

The big truck moved slowly ahead in fifth gear, but Phil had not gone five hundred yards before he was using the eighth gear. These low gears were hard to take at the start, though one did not realize until the truck had stopped what a screeching, grinding, jarring noise they made. For about three miles the going was slow;

soft, drifting sand with two bad creeks to cross. Archie, perched up on his racing camel, came part of the way with us and had a grin half-way across his face, for he had no trouble keeping up with us, even at the slow walk of the camel. We had to toil hard with the matting and shovels. Archie directed us to a crossing on the second creek and once we crossed the creek the going was fairly good. Phil was getting back into fourth gear and I said, "Thank you, Archie, old man," waving my hat; but with the noise of the truck he did not hear me. He was more interested in the capers of the car. It could not make the steep bank and Freddy had to give it a tow. As the car had only two men in it, when it came to any soft sand its back wheels would just spin until it dug itself in.

5

Lasseter Starts to Recognize the Landscape

AT FIRST we kept away from the hills. They were fairly steep here, and many gullies ran down. Later on in the day we encountered patches of mulga scrub. As the hills were only half the height of the morning ones we decided to try how the going was close up on the stony ground. After a couple of miles in as many hours, we came back onto the sand as the stony ground was spongy and soft, and we had to use the matting in many places. This meant we first of all had to throw all the stones out of our path, or the matting would have been torn to bits in five miles. So we dodged in amongst the scrubby trees and sometimes found a patch of good going. Then we came opposite to a fairly big break in the range. Mickey told us the spring was up there, so we turned and headed six or seven miles for the gap and found a nice place to camp.

Our camp was about three miles away from the spring. We arrived about two o'clock and all excepting George and Phil followed a brumby-pad to the water. There are many wild horses in these parts.

What a task it was for six of the men to fill the three nine-gallon drums with water and bring them back! We did not return until 5.30 and the brumby-pad was so narrow that we were compelled to walk in single file. This meant that each man alone had to carry a drum of water on his shoulder. It was hard-earned water; and it wasn't much better than Archie's. We were all very tired but George had cooked a good feed which cheered us up. We decided that rather than go through all this again, I would risk getting to Dashwood Soak, the place that Archie had told us about.

It was in this break of country that I saw wolfram, only a small seam about five or six feet in length. I took a sample and later found it was of a very high value. I told some wolfram miners later but never heard any more. I said that it looked only a bit of a squib; so, if they went up there, it probably was.

Some fun was caused that night. The captain and Lasseter had been trying to shoot something. It was very dark before they returned and, after they had about two pints of tea each, it was discovered that they had drunk the washing-up water. We were using Hamilton Downs water for the slopping, and they had put milk and sugar in it. The Englishman said, "Well, that beats cock-fighting!" He then took some of the tea made with spring water, milked and sugared it, tasted it and then threw it away and said between much spitting and spluttering, "The bullock soup is preferable to that." It *was* terrible, too, for it was half horse.

It was on this day that Lasseter started to "recognize" things. There was a big fault in the granite rock about six hundred feet up. Pointing to this he explained to us, "See that cave up there—I can remember that I camped in there on my way out to the west."

I looked at him in astonishment, and to see if he was doing any leg-pulling, but he seemed perfectly serious and I decided to hold my tongue. George grinned at me, and afterwards asked what I thought of the statement, since we both knew that nothing but an eagle ever camped up there. We let it pass, however, just in case it was a bit of leg-pulling.

But again that evening Lasseter went up to two bean-trees and, patting them, said, "This is where I rigged up my hammock, between these two old cobbers." This was his second statement that day of something that he recognized, but this time he picked trees that don't live much over seventeen to twenty years.

I went to bed that night with very troubled thoughts. Next morning while Phil was doing some repairs, Harry and I went across a big creek to see if the going was any better than the way we had come in. When we got away from the crowd I asked him what he meant by those statements. He at once flared up and said angrily, "You think you know everything!" When I told him the trees he picked were not more than fifteen years old, he declared that his trees were two exactly like these. "Well," I said, "since you believe that, tell me something of what is ahead today."

I did not know this part of the range. The Macdonnells merge into another range, and it seemed to me, by the changing character of the ground and scenery, that we were coming to the end of this part of the range. I thought it would be a good test as this was easily the most remarkable change of country that we had encountered. However Harry would not risk predicting what was ahead of us.

When we had started I found out that, even with direction and a good bushman like Colson, it was very hard to get through the maze of low hills and keep direction. All that day we were winding our way through these hillocks, easily the most remarkable rock formations we had come across. Harry was sulking because I had reminded him that he was not keeping to his contract, which was to tell the location of the reef, and I also made a few vigorous comments about his forgetting such striking country.

At last we came to the end of the hill country, and, according to directions, we were near the Dashwood Creek. This seemed likely by the fall of the ground as there appeared to be some hard going over a sandy patch, which in this country often shows that one is approaching a creek.

We decided that Colson and I would go ahead and try and locate this creek. I took my push-bike and waterbag and made for some higher country, to see if there was any sign of water from up there. Freddy Colson cut across the lower country, picked up some brumby-pads and followed them to where the soak was, a distance of about six or eight miles. At last I reached a point where I could see the creek, so I rode my bike across and met Freddy returning. He was all smiles and said, "Have a drink," for he, too, had taken a waterbag. It was about three hours since we had left the truck, and I had not taken a drink so I was fairly thirsty. When I had been drinking a while Fred said, "Hold on, don't drink the lot." Gee! it was grand to have a drink of real water, not the stuff we had been drinking, half bullock or horse.

By the time it had grown dark we were wondering if we would be able to locate the truck, for once we got to the little hills everything seemed the same. After a couple of miles we climbed on one of the rocky knobs and could see a powerful spotlight. Phil had had the foresight to rig the wireless-mast with the spotlight on top and some members of the party kept it moving from side to side covering the whole of the country in our direction. I suppose we were about four miles away when we saw the light. When we got in I thanked Phil, for it had been of great assistance.

Freddy told them of the lovely water that was in the soak, and of how much he had drunk. None of the party had washed that day so when the good news came, all hands got dishes out and had a much-needed wash.

I will say this for Captain Houston—he could rough it with the best—not a single complaint. The only member of the party to whine was Lasseter. He never missed an opportunity to remind me that I was not managing the water question very well, and saying that if it did not improve, he would have to reconsider his position.

George was a rock to me in these little passes. He would butt in and lead me away and warn me not to answer him, for he could see that Lasseter would make any old excuse to break up the party.

Now that we had unlimited water I told them that we would stop for two or three days and have a thorough overhaul and wash-up, and it was a happy crowd that sat around the fire that evening. George, Phil and even Harry burst into song sometimes, and when a real old-timer was being sung Freddy and Captain Houston would chip in. Up to this night Harry had not taken any part in our camp-fire parties. Of course, sometimes the yarns were rough, and, as the days went on, Harry showed a distinctly religious turn.

We were up very early next morning, everything went with a swing, each man to his job; jokes were cracked amid much laughter. A bar or two of song broke out; here and there was a slight pause when someone broke out with that ditty "The Girl I Left Behind". This was life and there was no earthly reason why the whole trip could not be conducted in such spirit.

I had a look around to see that nothing was left lying about, then, taking up my position, called "All aboard", and amid the screeching of the eighth and ninth gears I could hear George trying to get above it with "Off to Bonny Scotland".

In about thirty-five minutes we cut the wild horses' tracks or pads. We had at last left those puzzling hills. It was good going now. Phil got back to the second gear which had practically no noise and this was a relief. We came to a levelled patch of ground, where we stopped for a minute or two and all hands agreed that this hard, level ground would make a good aerodrome. There were only three or four trees and a few bushes to remove to make a very safe landing ground for a small plane.

I was asking Phil some questions about its possibilities when there started a chorus, "We want clean water! We want clean water!" so I said, "Let her go, Phil," and in a few minutes we were on the bank of a very fine creek, a big waterway, some three or four hundred feet wide, with grand, luscious-looking scrub and gumtrees. We arrived just before 8 a.m.

This Dashwood Creek was a very pleasant-looking place. Its damp sands were tossed up in all shapes for every time a stallion had led his brood into the water, he scraped the sand away with his front hooves. By his doing this the water was fresh and cool and it prevented little birds polluting the small waterholes.

We got the shovels out and levelled the sand and made a good crossing. When all the vehicles were across, the first order was to boil the billy, for none of us had had much breakfast. Some had not even had a cup of tea so we all sat round sipping the hot, delicious tea.

The morning still had a nip in it as I planned the day's work for each set of men. I told Harry to unpack the truck and offside for Phil who had many minor repairs to make that day. George and Freddy and I set about sinking the hole to make a permanent well. It was an easy task, for the three of us had often done this sort of work. The timber was handy and Mickey helped the captain cut and carry logs. In about two hours the soak was done and we lowered one of our packing cases so that full buckets could be dipped without stirring up the bottom. Next, all our kerosene tins and buckets were put alongside a big fire, for everyone was impatient to get his dirty clothes off. Six naked men were standing in a row lathering themselves with soap and pouring water on their heads. Each helped scrub another's back. It was great fun for Mickey, for this was his first glimpse of white men without clothes. Of course, there was much banter with him, but no amount of persuasion would induce Mickey to have a bath. Later on I got him to take off his flannel shirt and wash it, but he made no attempt to wash the top part of his body.

Only Phil and Harry now had not had a wash, so I asked Phil how long he would be using the oxy-welding plant. He replied very emphatically with his funny English style of using Australia's language. I told him it did not matter today, to let it go. I put on a couple of buckets of water on the fire and inside ten minutes Phil was having the time of his life. But Harry would not strip off; he only washed a change of clothes.

George and Freddy were feeling a bit peckish. They were cooking, and the captain was peeling spuds and making himself useful, as usual. After Phil had got into a pair of shorts, shoes and big hat, we got out the wireless plant. While he and Harry were unpacking it, Mickey climbed a very high tree with a small rope, which he fixed over a limb so we could pull up the aerial. It was a very creditable job to climb that tree, I am quite sure no other member of the party could have done it. I felt anxious when he was coming down, because his eyes were bad that day, but he came down just as easily as he went up.

After a good hot lunch we took it in turns to wash our dirty clothes, excepting Phil and Harry. They put in all the afternoon trying to fix the wireless so that we could send our first message to Sydney. Our calling time was at nine o'clock.

I drafted out a short message. I was determined to control all the messages and we were only allowed fifteen minutes. By dark two very disgusted men gave it up. The receiving set was linked up so that we would be able to hear the news of the day. Harry told me he thought he could fix the wireless, if left alone, as he had attended wireless lectures daily for three weeks before leaving Sydney, and Phil had had only three short lessons.

By this time George had prepared another big feed and it was a very happy party that night. That bum wireless set only had two pairs of earphones, and it was a great shock to all that no loudspeaker was attached, although Taylor and Lasseter had both insisted that there should be one. The English and Australians were playing a cricket test match that day and interest was keen. The captain took charge, and wrote down the scores and the descriptions of the game, and for a while even the gold reef "ten miles long and eight feet wide", was forgotten. With two Englishmen in the party the discussion was keen but after a time cricket records and "Don" got monotonous, so someone said, "What about a song?" George started and we all sang heartily except Harry and Mickey who were listening in. The wireless set was down at the soak, to get the water connection, so our noise did not matter. The night soon grew cold and we were all pretty tired, and glad to turn into clean beds. It was pleasant to tuck one's head under the blankets for they smelt as only clean-washed blankets can smell; not like the last few days when one could hardly stand his head on the lee side of himself. The bullock and horse water had stable charms of its own.

I had better explain what a soak is. When these waterways or creeks are in the making, big holes are washed out, and in nearly all soaks there is a bar of rock at the lower end that impounds the water. In hundreds of years these holes become filled with sand and it is this sand that protects the water from evaporation. Big soaks like this will last for years. Much good work could be done all through this country by placing underground bars on banks; good milled clay is as good as cement. It is always handy to the job and, of course, only costs pennies compared with the pounds of any other method. Date palms, nut trees and root crops could be planted along the banks of the waterholes; such things would look after themselves. Work like this would help pave the way to a better understanding with the natives and give them something for a change. This is the only way we will ever get to understand these children of God. Of course such areas should be reserves in the strict meaning of the word. Only qualified anthropologists should be allowed on these reserves, and then only to see that the plantations were doing well. All communication should be with the leaders of tribes or bands as recognized officials. The native laws and morals would not be affected and, in the course of time, some glimmer of understanding would come to these Aboriginal people.

This was a great bit of country with a big water supply; the waterhole was over four hundred yards long. Below the top of the sand it would last a big mob of cattle for months. This place was Archie's last stand if the rain did not come but this was on the edge of the "myall" country (that is what the natives of these parts are called who still live in their natural state). If Archie had to bring his cattle out here his losses would be great, since the bullocks are "fair game" to the owners of this country.

6

We Clear an Aerodrome

NEXT MORNING nobody stirred, for it was very cold. We waited until the sun had some warmth in it before we rose. Phil was able to complete his repair jobs, with George offsiding for him. Fred and Harry would mend the spare tubes and fix all the wheels; then Captain Houston, Mickey, Coote and I went over to clear and measure up the aerodrome, about half a mile away. It had one bad feature—there were steep hills on the southern end. We removed a few trees, and some old-man saltbush. This did not take very long, and there was a clear runway for five or six hundred yards on the west and east. Coote assured me that it was a safe drome, with enough room for a cross-wind landing or take-off, so the next thing to do was to put it on the map.

While we were doing this work I had asked the captain if he would accompany Lasseter when he took our bearings at noon, as it was very important that these should be correct. He consented, much to Harry's annoyance. I explained the position to him. There must not be any chance of a mistake, for I intended to give the position of the aerodrome over the air to the Sydney office. I had already explained that I was doubtful if Lasseter was capable of taking these bearings and I wanted the captain to check his work.

The captain had with him a very valuable book on navigation. It was lent to him by Rear-Admiral Evans, Commander of

H.M.A.S. *Australia*. After the shot had been taken the captain checked up with the aid of the book, and found that Lasseter's figures were all wrong. As I intended to push off early the next morning, no position was taken. The captain told me he was doubtful of Lasseter's ability, so I got the two of them together and tried to thrash it out. There was a good deal of heat in the argument, and when the captain quoted from a passage that indicated to his mind such a bearing could not be, Harry became sulky and would not finish the discussion.

The captain had lent Harry this book in Alice Springs and now Harry studied it at every opportunity. Lasseter had said at a Sydney meeting that he held an American diploma of surveying. This turned out to be one issued by a correspondence institute. He reminded the captain that he held such a diploma, and I was certainly in a quandary as to who was right. Then Lasseter went out of bounds and said all sorts of nasty things that had no bearing on the question.

Later that afternoon we had to re-pack the truck to make extra floor-room for the four forty-gallon drums of water. These were ones we had used our petrol from. The captain was helping and I told him to make a place for a box of ammunition right near the tail-board. He did this, but after a while Harry moved it back into the cabin. The captain saw him and said to Lasseter, "The Leader," (for that was how he addressed me when doing any work) "told me to put that box at the tail-board."

It looked as if there might be an argument, so I said, "That's right, Harry, I told him to remove it, as there was no room for it on the floor of the cabin."

Then Harry turned on me and accused me of hobnobbing with the aide-de-camp and asked me what would happen if we were "attacked" suddenly. I had to give him a lecture explaining that all the ammunition in that box was for sporting purposes, and until he found the reef, there was not much chance of using it as the turn-out was very expensive and time meant everything. I said, "If you follow my instructions on this trip, you will never need to pull a gun on a native. Another thing, I made a declaration that those arms were for sporting purposes only and I mean to see that that is all they are used for."

Then I got under his argument by saying, "Why are you behaving like a new-chum that has never been out here before? One thing that is very conspicuous in all your stories is that not once did you ever tell of any contact with natives, and here you are with three noisy motors and seven other men, trying to put up an argument that we may be 'ambushed'."

The others knew that something was happening so all stood away, but when I carried the ammunition box back to the tail of the truck all became busy and I felt sorry for the captain because his behaviour and manner of addressing the others had been all that could be desired in any sort of company.

There was another member of the party whose tongue was a lot too long, and how he came to get into the argument I cannot remember, so I shut him up by saying to him quietly, "You are not helping, you are only making matters worse." Then I appealed to him to stop the slinging-off and try to behave as a grown-up man should. I said, "Try the other stunt and agree and say 'That's right', even if you don't agree. You can see that Harry is half-mad, and for the peace of the party try to fall in with the others."

I then went over to Harry who was trying to get the wireless together, and we had a long talk. I put it to him in the same way and he readily agreed. Then I had to wait for an opportunity to tell the captain what I had done, so that, when he was suddenly treated with such unusual politeness, he would know how it all came about.

Until then the captain had ridden in the big Thornycroft, but after that he took up a position with Fred in his truck. I was sorry, for the ground work on the big truck was heavy and there was one man less to do it. The laying-down of the coconut-matting sounds easy, but it is extremely hard when there are three or four relays running, and each time the strip of fifty feet has to be rolled up and carried to the end of the other length that the truck is on. Sometimes the driver would get so much way on that he would get over the sandy patch without stopping. Then he would pull up on good ground and wait there, which was sometimes one or two hundred yards further on.

George and I often talked of how the captain worked on this matting, for it was the dirtiest kind of job. Of course one could not help sweating and you can imagine what it would be like to have thorns and sand running down under your shirt. He didn't even complain when we ran into a patch of porcupine spinifex. The sharp spear-points of this are as bad as black-thistle thorns and for him to do this just "for experience" often amused George and me. It was men of the same breed, and maybe the same school, who did so much exploring and early pioneering work in this country.

I must not forget our Australian Aboriginal, for he was just as good as any of the whites in our party. Mickey did not want any telling when the matting was out; he always worked one strip and he was a very strong man. He used to show signs of pain when one of the cruel cat head thorns went into his hand. One simply had to get it out for it burnt like a red-hot needle. As Mick's eyes were not good enough to get them out, he would come either to George or me.

When there were four men on the matting it was not so bad but later on when it got down to just George and me and with the hot weather on us, I can tell you we often got the trembles. That was not all the trouble either, for often the truck had to be dug out with shovels. It was set too low for this kind of work, with only nine inches clearance. It would gather roly-poly grass and other things which would roll up into a bundle until they stopped the truck. Sometimes the matting would get mixed up with the bundle, and it often took thirty minutes to clear the undercarriage.

Spinifex was our best country, because this tough tussock grass provided a good grip for the wheels. This grass is almost useless for stock fodder because of the hundreds of sharp spines. These have given it the nickname of "porcupine grass". Stock will eat the seeds only and, when let out with plenty of green feed, will sometimes forage in amongst the spinifex for a bit of dry feed. It grows only on sandy country, fortunately, a bit of rain ruins it; but if it could stand rain, it would be a terrible pest over surrounding good country. It is very inflammable and great care has to be exercised with fires as two-thirds of each clump is composed of old dead grass. It burns so fiercely that a man on foot has no hope of getting out of its way, but must burn a break to escape it.

Our second night at the Dashwood was very pleasant. I think I was the only discord for I had examined that bum wireless plant. It had a board with over a hundred holes in it. Each hole was supposed to have a plug, each hole and plug were numbered, not straight-out numeral numbers, but things like 1A or KE or G to I. Then each plug was connected to the copper wiring; the company was sold a pup right enough.

After we got the cricket scores and the news of the day through, the earphones were handed over to Mickey and Harry, but at night there was a light wind and no matter where one got, some part of him was cold. The yarn-telling party soon broke up; bed was the warmest place.

I had taken the precaution to boil a drum of water for drinking purposes. This was my third expedition into the sandy plains country and I knew the value of boiling all drinking water. I had experienced the ravages of dysentery and to my way of thinking it will wreck a man quicker than any complaint I know of.

Everything went fine in the morning and all were anxious to

move ahead but I noticed that no one was game enough to take a bath before departing from the bonzer water. It was biting cold and all had their top-coats on. The start was easy, and for five miles or so the going was good, but we soon were out matting and breaking down the banks of small creeks. As the day wore on we became slower. We ran across a patch of gibbers; that is, stony ground, with soft swelling ground underneath. This was painful, for every stone had to be taken from under the matting and, of course, those on the big truck got the brunt of the work. The Chev. truck and car had no matting and thus often had difficulty in getting a start, so they were seldom on hand.

I think we covered only twenty-three miles that day, not quite two and a half miles per hour, but Phil and the captain had their experience of a big mob of kangaroos. Phil wanted to try his rifle so a halt was made while the slaughter went on. I kept one to eat for I am fond of kangaroo steak. Errol and the captain were on for a bit of sport, and wanted to make the place a camp for the night, but for the life of me I cannot see where the sport is in just shooting for shooting's sake. If the meat or skins were of any use it would be entirely different, so I said, "No, we have enough meat, so on with the game."

About 4.30 we ran on to a bit of hard ground with plenty of wood about. George said, "What about a camp? You know there is baking to do tonight." Twenty-three miles did not seem to be very far but there looked to be sand ahead and I can tell you it is no joke to camp on the soft sand. It gets into everything, and it is always covered in thorns or prickles. Besides, we were all very tired. Cutting down the banks of those rivers is not light work. It had to be done properly; Fred would not have anyone with him when he was running up supplies so we were making a track that would give him the least trouble.

I cut up a lot of kangaroo steak and fried it with onions; everyone voted it good. This is where I watched Harry closely and he did exactly as the others who had not eaten the hopper meat. They took the tiniest bit to taste, then discovered that it was better than a lot of so-called rump steak they had eaten. I felt pretty sure that this was Harry's first feed of 'roo although he told tales of how he had lived on it for months.

I was called hard names for not having a shoot but I explained that in the planning of the stores for this trip, all that had been gone into. I pointed out that the 'roo meat we had got that day cost the company about 3s. 6d. per pound. I explained that we had started on this trip fully six weeks too late, and that time meant everything, for we would have the hot weather on us before we knew where we were.

Mickey came in for a dainty morsel that day. There was a young joey in the pouch of the kangaroo which was killed. He cooked this in his own native fashion by burying it under the coals. By the way, this was extra, for he cooked it after his first steak and onions. In almost the whole of Australia this class of food belongs to the very oldest people in the tribe, and no one else is allowed to eat it. It was probably the first time Mickey had tasted it.

Next day, after the first five or six miles we found much better country. About mid-day we came to a very big gum creek with steep banks. We looked for three miles each way but could not find anything better. This was a dry creek. We were compelled to put in a cutting sixty yards long. While Colson, George and the captain and myself used the pick and shovel, the others cut timber and corduroyed the top of the sand. In about two hours we were ready. The bank on the other side was steep so we laid all the leaves from the tree-tops on the logs, then threw a little sand on them to bed them down, then all the matting on top. This would give the truck a better chance of making speed enough to clear the steep bank, so I told Colson to go across first.

He almost got to the top of the bank when something went wrong. Of course all hands were there to help, but it did not take Fred and Phil long to find out that he had stripped the pinion. We quickly unloaded the truck and pushed it out of the way. Then the old Thorny came across, quite an easy matter. It could have been as steep again without troubling it. By this time Phil had mastered every movement of this six-wheeler and it was here that he demonstrated its powers, for this was a tricky place with a bend as soon as it was clear of the bank. Phil and Fred were soon on the job and in ten minutes they had the broken piece of the vehicle out. It was a stripped pinion all right, and we had no spare part.

We held a hasty council and as the captain's time was nearly up and he would have to return, it was decided that Fred and Errol would go back to Alice Springs in the car, and Errol would bring the plane out with Fred and the spare part.

So Errol, Phil and I inspected the ground just nearby, and decided that by moving twenty or thirty trees, a very respectable drome could be made. As I had heard so much about what the plane could do I was anxious to have it closer up.

When the captain began collecting his things the old snarling temper of Lasseter returned. Lasseter would not give the captain the book on navigation. Lasseter had locked it away in his big tin trunk and no argument would shift him; he just sat there and sulked.

The captain explained that Rear-Admiral Evans valued the book very much. Then Harry started arguing that if he had known the captain was taking the book back, he would have brought his own, and we all knew this to be untrue, for until the captain had shown him the book he had never even thought of such a thing.

Then he clinched matters, for he stubbornly refused to go any further unless he had a book on navigation. Lasseter had been waiting for such a chance to show his authority. It was just sheer cussedness, for I will lay on oath that Lasseter never opened that book again. He did not want it, in fact he could not understand it.

Why Lasseter took such a dislike to this very likeable man I could never understand, for at no part of the trip had he been a passenger. In fact, it was actually through his help that we were so far ahead. I had a confidential talk with the captain and told him I was quite prepared to have a showdown, whether it ended the expedition or not. He would not hear of this, so I gave him my word of honour that I would return the book myself to Admiral Evans's own hands. The captain had explained to me that the book was out of print and was practically priceless. This copy had been with the Scott Expedition when Captain Robert Scott and others lost their lives, and because of that it had a special historic value. Rear-Admiral Evans was a member of that expedition. Later on I will give an account of where that book took me.

So the Chevrolet car was turned round and after a hasty lunch the three men, the captain, Errol Coote and Fred Colson were soon out of sight, on their way to Alice Springs, which Freddy hoped to make in three days.

7

The Plane Arrives

I THOUGHT that now the captain and Coote were out of the way Harry would be easier to manage. He was, in fact; he went out of his way to make things pleasant. He worked really hard clearing those trees. As a matter of fact he always did his bit, and was never far behind everyone else.

The surprise of the party was Mickey. He proved to be a good axeman. I showed him what was wanted; the trees had to be cut eight or nine inches below the surface. It was a brute of a job getting down these umbrella mulgas. On each tree there were up to a dozen branches that stooled out, and before we could get at the stump, several of these branches had to be cut away. They were seldom thicker than a man's arm but this mulga wood is tough and hard, one of Australia's hardest woods, and probably the most beautifully-grained wood that we have. It never shrinks or cracks and takes polish like metal.

It was Mickey's brain that solved the problem of getting these trees out. He dug a hole right under the branches, so that he could cut the stump below the surface. He was soon many trees ahead, so we all copied his method which made the work much easier and twice as fast. George and I looked upon this as a great win for the Aboriginal.

We at first cleared a big cross three or four hundred yards long, east to west on the first day, and south and north on the second day. Phil said, "The plane can come down on this," so we eased up then and widened the runways at a more leisurely pace.

I don't remember how many days it was before the plane came but it must have been four or five, for we had rested one day and George and I had prospected the creek both ways for a possible supply of water. I was told that there were some rock-holes some four or five miles further up, but I was doubtful, as there were no wild horse-pads anywhere near this creek. Also we did not see any small bird life.

These small birds are good indicators, because they are seldom more than six or eight miles from water, and one has only to follow their flight to know the direction of the waterholes. This is fairly easy, since they are always in big flocks and they do not fly far at a time. There are several kinds and they all belong to the finch type. The different kinds of parrots are also a good guide; they fly in flocks and very fast. They fly long distances, and it is only late in the evenings or early in the mornings that they can be taken as a guide, as they always camp near water.

We saw none of these signs at "Taylor's Creek". I suppose this creek has a name but our maps did not show it, so we called it after Phil. It was of considerable size and flood marks showed that at times there were twenty feet of water flowing down its channel.

The day we landed here George prepared tea and he asked Mickey how the kangaroo was holding out. We were all tired of it, and had given the remaining leg to Mickey. He replied, "Him no good, him 'tinkin' now." George smiled and said, "Mickey has got tired of the 'roo and thrown it away." But Mickey heard him and said, pointing to a bag of grass, "Him in that bag." George had a look, for the bag was part of his mattress half-filled with dry grass. Evidently the captain had had a joke with George that morning, but because of the breakdown he was not there to hear what George had to say. The rest of us thoroughly enjoyed the joke.

Mickey had only two feeds of 'roo, cooked in his own fashion. He was a station boy and not used to this kind of fare because Archie Giles gave all his natives the same food that he had himself.

Later on, when we had to cut a big gumtree down, there were two possums in it. Someone drew Mickey's attention to them but he was not having any. Ham and eggs, tinned fruit, tinned meat tongues and cheese were good enough for him. Anyway he worked as well as any of us, so was entitled to the best; he was the cheapest man for the company, working for his tucker only.

In the afternoon of the seventh day we heard the drone of the plane and inside ten minutes it landed, though not without incident. We all agreed that the runway was too short. The plane ran past the clearing by fifty yards and looked like running into a tree. By a stroke of luck it straddled a number of old rabbit warrens. Errol Coote, the pilot, said that he had seen them, so he got the credit for a very skilful landing.

Poor Freddy! I will never forget the look of him, he was blue and had been sick. They had had a very bumpy ride in the many air-pockets they encountered as they followed the mountains all the way.

They had, of course, brought the spare part for the truck, and also the mail, which contained some more instructions from the Sydney office. These again proved the office's pitiful ignorance of our work and conditions.

Anyway, very little time had been lost, and another emergency landing ground had been made and we now had the plane close up. I think this was about a hundred miles from Ilbilba which we intended to make our base.

Phil and Fred got to work straight away on the truck, and inside the hour it was fixed, so they joined us in lengthening the runways. Next morning Phil had a talk with Errol and they decided to test out the take-off with the plane. Errol got up all right, but I had an unpleasant thrill for it looked as though he was going to crash into a high gumtree. He had to take off towards the creek but he did a wonderful thing; he pulled the plane almost perpendicular, but hit the top of the tree hard. As he circled we could see a small patch of fabric flapping so Phil signalled him to come down. He came down again OK but again over-ran the runway. This meant more work and we levelled the ground up to the creek bank. Then we cut down several big trees, including the one he had brushed.

Phil patched up several breaks in the fabric, then gave the engine a final overhauling, and put the covers on. The wings were closed, and it was tied down. As both trucks were loaded and waiting, we moved off. It looked strange to see a beautiful painted thing like the *Golden Quest* tied up to trees, and we were leaving it to take its chances with the weather.

After we had got about a mile away we came upon fairly open country and we all agreed that this was where the drome should have been made, for by removing about a dozen trees, a drome a thousand yards long and six hundred yards wide could have been made and all that work avoided. I made the suggestion that we should clear this patch too for it would only take a couple of hours, but both Phil and Errol assured me that there was nothing wrong with the other one, now that the high trees had been cut down.

I did not need overmuch convincing, for this breakdown delay had meant inroads on our water supply. However, we were approaching the spring belt so I was not worrying. At any sign of a shortage of water we could man the auxiliary truck and go to one of these springs.

This country is almost free from sand, but the mulga umbrella-trees were often in our way and we had to cut them down. We made excellent time bowling across the flat claypans in top gear. The axe-work was nothing much, so all hands were having a quiet holiday, and it was a relief to get off the matting work. We had a splendid landmark all day—Haast Bluff—a peculiar mountain that seemed to stand on its own, for it covered the view of the hill behind it. It was easy to keep our direction, but all the same our track that day was a winding one for we had many clumps of trees to dodge. Although it was in a zig-zag fashion we made good progress.

We had covered forty to fifty miles when the Derwent, a big creek with a very steep bank on one side, came in sight, and this ended our hopes of a record run. The approach was bad, for there were three low sandhills lying almost parallel with the creek. We had to cross these on matting at about a thirty-four degrees angle.

This crossing took about four hours to accomplish. The cutting at the lower end was fully four feet deep, but unlike the other creeks, it had a good strong bottom, so once the cutting was finished it was only a matter of driving across.

Fred went up the creek two or three miles, while George and I prospected downstream; but I soon saw that if there was water handy it would be upstream. Fred reported that he had come across well-worn horse-pads and thought the water would be about five miles above the crossing.

I don't think any of us will forget the sunset that evening. It set right behind the Bluff which stood somewhere near 2,000 feet above the plain, only about a mile away. I am sure that even in Egypt there is not a more wonderful sight. The sun's rays seemed to be broken by shadows somewhere behind the mountain, giving effects of golden-hued lights and radiant rose-pink flames with dark fan-shaped shafts of shadow forming into the effect of a high point behind. After the sun had set we became interested in the earth's shadow reflected in the sky. It travelled rapidly and when the last curves of it sank over the western rim, it was suddenly dark. This country is renowned for the earth's shadow that heralds the coming of darkness. The spectacle occurs only in the winter months; the vertical heat rays going up off the heated ground in summer cause visibility to be only a few hundred feet.

We had a nice place to camp, hard level ground, and had the

water been a little closer we would have all enjoyed a bath; but gaining time was my main thought, so I would not go off our bearings. I don't mind saying that this decision was not very popular with the others, but Colson knew how valuable each day was becoming. The one big thing I did want to avoid was being forced to abandon the expedition because of the heat.

It was a busy camp that evening. We had many punctures; the two spare wheels and four spare tubes all needed mending. Our spare wheels with the tyres on were about one and a half hundredweight; it was hard work for two men to handle them. The mulga sticks that went through the two and a half inches of covers had to be bored out with a brace-and-bit. By the time the repairs and the cooking were finished we were all ready to turn in. From before seven in the morning until after nine at night makes a long day.

It appeared that Coote had told Taylor, when enlisting him, that the trip would be a real "joy-ride". Tonight Phil reminded him of it and asked when the joy-riding business started. He said, "Ask Fred Blakeley, or insist on camping earlier." George agreed with him and a quiet discussion took place, George pointing out that the long hours and hard work would crack up the team. I agreed to try to camp a little earlier.

8

Sand, Claypan and Mulga

NEXT MORNING was just perfect—nippy, but more bracing than cold. Phil and Harry staged a wild-west turn that caused much amusement and banter for days.

There were two channels leading out onto the open between the sand-ridges, so I told the drivers that I would run up one channel and see if it went through and I told Freddy to watch, then, if it looked all right, I would signal for him to come on. But Phil, who was behind him, misunderstood, and when I signalled he moved up the second channel. I ran across and waved my hat for him to stop, but he only waved his hand to me and gaily kept on going as he was in top gear and had his foot down hard on the accelerator. I ran quickly across and managed to intercept Colson who had taken the first channel. Harry was riding on the back of this truck, because he thought the big truck would come the same way. We had to change our course and chase Phil; by that time he was two miles away and going like the wind. The cover was as smooth as a billiard table and almost as level, but the channel that Phil had taken was not in the right direction. In about five miles we caught up.

We called out but of course he could not hear; he only waved his hand and kept on going. The channel was too narrow for us to pass him, so Harry started firing shots from his revolver to attract more attention. George saw and heard and advised Phil to pull up as Lasseter was working his artillery. When he pulled up of course I went crook properly, and Phil was the most surprised man in the world.

I said, "Where the hell do you think you're going?" He said, "Why, isn't this the way?" I replied in suitable language. So he had the pleasure of turning around on good, hard ground and had another fly-back at twenty-five miles per hour. It did not take long. It was too funny to be annoying, in fact we all enjoyed the joke for the whole trip—"the day when Phil cleared out!". Later Phil said, "My word, you spoke rough to me, as though you meant it, too."

I only grinned and said, "Where did you think Harry and I were? You know that our position is in your truck!" He only laughed at that and said, "I was so interested in the good going that I forgot all about you two. It must have been this crisp, clear mountain air that got into my blood."

The going alongside the Bluff was slow. By ten o'clock we had run into thick mulga scrub. This umbrella mulga was two and three times bigger than those we cut at Taylor's Creek. The branches when cut off were more than one man could drag aside. We had done only eleven miles by twelve o'clock. Of course I did not reckon in the runaway mileage.

We camped at 4.30 doing only five miles for the afternoon. We would clear about a mile at a time and this would take one to two hours. All hands were working and both drivers were up with the clearing gang. Harry got a bit nervy about the trucks being so far behind, so I told them to keep the truck up closer. Harry's arguments were sound; not that I was afraid of natives, but in case of fire.

At 4.30 there was a nice open pocket and George suggested it as a good camping place. I had agreed the night before to camp earlier and I knew that as the last spares had been used there would have to be much repair work. When Phil pulled up George said, "Pull her out of gear, we are camping here." Phil could hardly believe the good news, and he called out to me, "Is that right?" I nodded. He said, "Gee, what's the big idea in aid of?" George laughed and said, "You can do all your repairs in the daylight." Fred Colson joined in with, "Just as well—you have a flat tyre behind."

But it was a far better arrangement. We had all the repair work done before dark and, as there was no damper to cook, we had an evening of leisure and were able to enjoy a game of cards.

Early that morning we had passed within three miles of a big granite spring. We did not water there because we couldn't get the truck close to it and we thought the water would hold out. This spring was really Colson's country. He had paid some money down on it but this was the first time that he had seen it. Fred had the idea that the wild horses that ran on this country would be worth getting; he thought there were possibilities of taking three or four hundred horses off it a year. After seeing the type of country, however, he reckoned he would break a bank to try and run a herd of horses. The spring belt in this country was so extensive that it would take an army of men to hold the mob. I wonder what those city guys would think of their pet terms for this country—"The Dead Heart", "No Man's Land"—if they saw this gusher spring putting out 50,000 gallons per day. The stream goes underground only about half a mile from the spring. The spring comes up out of the granite. I know that it is not true granite as there is a deposit of limestone covering it. But this spring has great possibilities; the water gushes out with considerable force—I would estimate at about fifty pounds pressure.

After carefully scanning the natural fall of the watershed of this country, I would not be surprised to find that the stream could be taken two hundred miles by pipe-line. I would like to make all this country a natural reserve for the natives who own it, and have the tribal boundaries surveyed so that their white destroyer could not interfere with them.

Next morning Fred and I went ahead to see if we could find a break in the timber. We came to taller trees and climbed one and, as far as the eye could see, there were no open pockets or lanes. We went back to the men on the trucks and informed them that there was nothing else for it but to cut a track straight ahead, expecting that the alternating of types of trees would assist us, for the tall timber was more open.

Six axes set to work but by lunch-time we had gained only three miles. We had a good land-mark, Mt Liebig, some 3,000 feet high, which enabled us to keep direction easily. This tree-cutting was hard work. Then Fred, George, Phil and I had a talk about the possibilities of strengthening the bumper bars of the big truck so that we could drive straight at the trees and smash them down, and we decided to give it a go.

I had noticed a gidgee or myall-tree. This kind of wood is tough and wiry and hard, so we cut a stout stick and fashioned it to fit. This is where Phil showed his training as an engineer, for he carried supports seven or eight feet back on to the chassis, then cross-stayed it underneath. We then tried out the idea. It hit the first five or six trees, knocking them flying or breaking them off.

All hands, except the driver, were pulling the trees to one side. This was good; for we were making headway. It soon got very tiring, however, so we all got aboard and I told Phil to get as much speed on as possible and hit them and roll them over to crush them down as the truck forged ahead.

I said, "I will watch thirty or forty yards ahead and pick out the line of least resistance and you watch close up." So between the two of us we found a way through. Phil had a fulltime job watching just in front and at the same time keeping an eye on my hand to get the direction. At first we stopped to cut down the big trees that we could not avoid and after a little while we got a run of about fifty yards with no trees excepting one which prevented us getting into another opening, so I said to Phil, "Drive straight at it and hit hard." He did and knocked the tree out by the roots. He hit it so hard that it fell clear of the truck. This was a great discovery so we started hitting everything before us. It was wild work, for sticks and boughs often came through the lowered windscreens. The cab of the truck was a strong affair and it got many hard bumps in that run of an hour.

Phil said, "I will have to stop, something is wrong with the radiator." It was boiling a treat and as we were in a fairly open place we decided to call it a day. We cut down a few trees and made a really good camp. Phil examined the radiator and found that it was blocked up with furry stuff, the flower of a plant called pussycat tail, which took a long time to clean out. Mickey was given the job as there were many punctures to mend.

With all our bumping and hard work we had done only twelve miles, but there had been many delays, first changing a wheel, and later mending punctures. We were all in much better heart now, for this bumping racket was exciting and the thrill of it lasted all the evening. Everyone had tales to tell and there was much laughter and fun.

Next morning Phil decided to remove the headlamps. George gave a yell when he saw this and said, "You chaps look out today. Phil is clearing the decks for action, there will be something doing today." And sure enough there was; we met our Waterloo just before lunch. It was a tree about nine inches thick which withstood our onslaught. The sudden jar threw us all up against the windscreen and gave us quite a shock. It broke both the side-braces and new ones had to be cut. We then discovered that both front wheels were flat, so between the repairing of braces and mending of punctures we overhauled the outfit and made a fresh start.

This stop and lunchtime cost us two and a half hours, but it was great fun after that. Harry would see us heading for a tree that looked a bit big and yell, "Stop her." I nearly died from laughing later when there were two trees and George yelled out, "Stop, stop, there are two of the blighters," but my hand was still out so Phil could hit them. They went down but it stopped us. We had to back out to get them clear of the truck. George roared, "Spare me days, are you deaf and blind or what? Didn't you hear me calling out, 'Stop her'?"

By this time Fred had pulled up and he and Errol were giving a hand. We all laughed at George who was pulling the stays off the cab and saying, "No more doubles." This had shaken the truck for I can tell you that when six or seven tons of truck hit those trees it jarred to the core.

While we were going through that timber belt no one stole forty winks. Often Mickey or Harry dropped to the floor; they were sitting on the off-side and got more than their share of boughs and sticks as we had all the glass down. The glass that had been a fixture had long since disappeared and we had nailed boards in these narrow panels. I will say this much for Harry; he never asked for an alteration, although at times the danger was very real. When a branch was pointing straight out at the truck, there was not much time to dodge.

When we were moving we made good progress, but the repair-delays knocked off the profit. It took us five days to go through that scrub of only sixty miles—the pussycat tail gave us no end of trouble. Phil got a brainwave and burnt them out of the radiator with the blowlamp. After he had done it two or three times I asked him how much more would the radiator stand. He agreed that the operation was risky. As we had the makings of a dry-blower aboard, we decided to rig up one.

A dry-blower, used by miners, is an arrangement of sieves and screens. They toss the dry earth into the blower, at the same time shaking the screens. The common method is to depend on the natural forces of the wind, but when working alluvial, they use a fan to create a force, thus blowing all the dust clear while the heavy material drops into the bin.

We had three sheets of expanded metal and one of these was cut to make a screen seven or eight inches in front of the radiator. This prevented the fluff from getting into the holes of the radiator, but of course as it was in the line of action it often got knocked. While we were in the timber, it came in for repair pretty often.

On our seventh day of the timber belt, we were making better progress. The radiator had been fixed and we were getting more expert at hitting trees, also dodging them, so we travelled fifteen miles that day; but the punctures were increasing. As we moved into the open, the undergrowth became thicker. It was these little sticks that were going through the tyres. About four o'clock we had to stop for puncture repairs so we called it a day. We pulled up in an opening with tall trees all round and they made a fine windbreak.

We had a lot to do; taking off those tyres was getting harder each time. It often took two men ten to twenty minutes to get the rings off as they were getting very rough usage. Their many bumps from trees had put kinks in them. We had one misfit tyre. It was a brute of a thing, and whoever had the task of taking it off always got hot and bothered and much cursing was let loose. It was this tyre which nearly settled Phil. He was holding the tyre spanner and I was using a double-handled hammer and was hitting the spanner trying to force off the ring. When the spanner slipped it came down on his thumb with a glancing blow. I thought his thumb was off at first. It looked a very bad smash. That ended the work for the evening.

The first-aid chest was brought out and from it Phil was given a dose of brandy. While he was sipping it he winked at George and said, "I got two inches of brandy in a half-pint pannikin for a hurt thumb, I wonder how big a hurt it would have to be to get four inches." The thumb was too ugly to investigate to see how badly it was smashed. We put the splints on and fixed up his bed. We had our tea and after a while I asked him how it felt. He said, "Pretty good, just as though you had hit it with a four-pound hammer." So I gave him another two inches and that put him into dreamland.

We all thought it was a case of stopping for a day or two. The evening flight of parrots showed that there was a spring in a gum creek, only about three miles away, so we felt all right as far as water was concerned. Next morning Phil said, "I think I will be able to manage, but won't be able to do any repair work."

We were quite satisfied to do anything as long as we could move ahead. We all admired Phil's spirit for he had taken a thumping whack, but never said a word although his thumb was in splints for two weeks.

Since there was still much work to be done on the tyres and tubes Fred and I went ahead for four or five miles. It looked as though we could skim up near the mountain and dodge a lot of timber. But the ground altered to small creeks and spongy, stony ground, so we were forced to keep in the timber country. Although the timber was easy to get through, we often got into trouble with the soft patches, and mostly with the front wheels. This always meant backing out and trying another place.

9

Smoke-signals

WE WERE NOW climbing higher and our course led us past the splendid Mt Liebig. I think it is the most spectacular mountain that we have in Australia. It makes a grand sight as it stands out so commandingly from the surrounding landscape. Our map showed it to be nearly 4,000 feet high. It is composed of hard basalt flint, whitey grey in colour.

There were three folds that resembled the lip of a volcano. The first fold was about 1,500 feet up, then a sharper rise led to the second fold 1,000 feet higher; then came the ragged edges of the top. Even when one was on it, the mountain still resembled the outside cone of an old volcano. The south side looked to be almost perpendicular, but we did not get a fair look at it. When we got up to 2,000 feet we began to understand what this water springs belt meant, for we could see many open waterholes dotted here and there out on the plains, and in the distance there appeared to be many wild horses.

There was also another fine view, Blanche Towers, as we went towards Ilbilba. The Towers showed over 4,000 feet on our map. There were three peaks, two about the same height with a shorter one in the centre which appeared to have many sharp peaks at its base. We were all the time about fifty miles away, but in the distance it looked like three giant stalactites sticking up in the air.

I have heard that these peaks are too steep to climb, but from a distance they are a fine sight. They and the Liebig are quite worth the journey to see. I have often seen photos of Ayers Rock, but have not seen any of these two outstanding mountains. My own photos of them taken on this trip were accidentally burnt.

We were late starting out that morning, and the going was slow, but a most remarkable sight appeared just after we started. From every direction we could see smoke-signals going up. Before we started there was not a sign of one, but as soon as the old Thorny got into its low gears, up they went. By the number of signals I should say the native population was about 3,000.

That day we made good progress. We traversed many large dry swamps and although they were rough and bumpy the going was good. We saw much evidence of wild horses. In the afternoon we were all on the lookout for an old camel-pad, a track cut by Bob Buck who had had the contract to take petrol to Ilbilba where the Mackay Flying Expedition had made its base the year before. The pad would lead us to the water and Ilbilba. However, a camel-pad a year old is not an easy thing to find in that moving sandy country.

We had Mt Tudor in front of us and this made it easy to keep direction. When we got within two miles of the Mount, a halt was called, for the country ahead was very broken. We had lunch and decided that while the washing and repairs were being done, I would go towards the mountain. The purpose of this was to see if I could gain enough elevation there to survey the landscape. Coote came too and there was a sheer cliff of over 1,000 feet facing us. We went up as far as we could but not high enough to enable us to get a view of the lay of the sandhills. We could see they were numerous; we would have very sandy going. I thought I could pick the place the camel-pad should be.

We were away about two hours and when we were on our way back, I got a bit bushed and could not locate the truck, so Errol fired a shot. We were only three hundred yards away. They had turned the trucks, for Fred had gone back half a mile and fortunately had found the old camel-pad. Now there was no doubt that we would find Ilbilba.

There was a new spirit in our movements. Even an old camel-pad seemed Heaven after the last weeks of gruelling travelling through the dense scrub. Phil soon found that it assisted us to keep one set of wheels on the pad and it enabled us to get over many sand patches that would otherwise have meant laying the matting. But the country had changed; we were following a line of low hills and even the sand had changed.

This new stuff was soft and silky, and for a while I was beaten as to how to cope with it, because when we laid the matting on it the wheels would just spin round. The grip of the wheels on the matting brought it back so that we had the matting moving instead of the truck. It was hard to believe at first, but when it happened the second time things looked serious. Here was something I had not struck before. In fact when the truck ran into this sand, it threaded the matting back behind the truck like lightning.

We were going up a slight grade, and it was the front wheels which were doing the damage. Phil could not keep them straight and would twist over until the front wheels locked. There was nothing else for it but to dig two trenches in order that Phil could keep them straight. We did this, because I was afraid to risk any more pulling through with the matting. Everyone could see that it would soon be torn to bits. We then laid bushes and grass in the trenches and the old Thorny went up without the slightest bother. What a relief it was! But the most peculiar thing of all was that it was worse going downhill. The front wheels simply could not be kept straight, so we had to dig more trenches about nine inches deep and fill them with anything we could find as there was no spinifex grass here.

We next ran on to a level bit of hard ground with plenty of wood handy, so George called "camp". I was quite satisfied, for these new tricks of nature took some thinking over. This was something I had never heard of before; sand that matting could not be laid on! I am afraid I used a lot of bad language over this last mile, but it only seemed to amuse the others. They were all chuckling while camp was being made.

I was not doing a thing. Just standing looking at the country ahead. There was a similar hill in front and I was thinking: if we have to trench everything these sixty miles to Ilbilba will take a week. We had run about five and a half miles in forty minutes, the last half mile had taken two and a half hours, so while tea was being prepared I walked ahead and as far as I could see the country was all the same. This trenching, and laying grass and tree-tops in it, was a terrible thing. It wasn't such hard work actually but slow and tiresome, for the filling-in material took a lot of gathering. While I was standing on the next rise, I had an idea. The driving wheels were not digging in but the front wheels were. I had seen them pushed yards while on full lock, pushing a big bank of sand with them. I wondered if we could make spoon drains just deep enough for Phil to control them and not put in any filling. The idea seemed worth a trial so I went back and had a feed. It was a feed I liked. Our big spell at lunchtime allowed time to boil a pot of spuds. George had mashed these with a few raw onions cut up fine and three tins of bully beef. This dry hash was always good and easy to prepare so while we were all tucking into it I told them what I intended to try next day. It was a happy camp that night, there were no punctures to mend—that had been done at lunchtime.

It was a beautiful night, not a breath of wind, and there was some comfort in playing a game of cards, and the dry wood threw out its warmth many yards. All through the evening there was much chuckling over the half mile in two and a half hours, especially when Phil tried to string some of my swear-words together. He would get them hopelessly mixed up.

Anyway, the cursing must have done some good, for next morning the two spoon drains were tried and the old truck went up quite easily and in fourth gear at that. We made a lesser drain downhill and we had no more bother with this silky sand.

We struck a new demon that afternoon however, drifting sand among big boulders. This was a clean knockout, for these boulders simply had to be removed. It was tough work, but, thank goodness, there was only about a mile of it. That mile took three hours. Then came a great run of scalded flat for six or seven miles.

"Off to Bonny Scotland, riding down the Clyde" was being sung by Phil and George. This was great, but it ended in a stony hill, easy to climb on our side, but the other side looked terrible: all tossed-up sand-drift, not in folds, but like a mixture of pyramids. It took some time to survey this job, for in this track-making we had to think of coming back, so we put a bend in the road and found an easy way down, but we still had to pass through three or four hundred yards of this pyramid of tossed-up sand. It was not so terrible to cross as it looked. We levelled the top off and laid bushes and grass and it made a splendid road. This was the last of the undulating country—we could hardly call it hilly.

We now ran between the sandhills and from this slight elevation of the hill could see four or five miles ahead. As soon as we got on to a clean bit of hard ground George called, "Campo! Good wood here." I was impatient to go on, but days ago I agreed to let the others call "campo", so, instead of helping make camp, I went ahead to see if there was any chance of crossing over into other channels because this one was gradually pushing us away from Ilbilba.

About three miles up I found a crossing, but it was a matting job. I then saw a break in the next flute. These short breaks pleased me, for what happened up to the first break would happen again at the same regular intervals as long as this particular system of sandhills lasted. I had had previous experience in these long lines of sandhills and knew that if a break occurred in five miles you could bet that it would show another break at that distance in the next channel.

From here on, we spent many weeks threading our way through this long line of sand-ridges. They always run nearly true east and west from a hundred to four hundred yards apart, leaving long lines of wind-swept flat channels in between. It was necessary for me to walk one of these channels and find the system of breaks before taking the truck up. Sometimes it was four to five miles before a break occurred, but seven times out of ten the two sandhills we were travelling would close up into a V-shape. Just before this would occur, there would be a break in one or other of the sand-ridges and very often we would get through without matting. Just how these breaks occur is a mystery and, I think, will always remain so for the country is always the same; flat and level with these sand-ridges sometimes running for hundreds of miles, always keeping the same direction and about the same height—forty to a hundred feet—not very wide, a hundred and fifty yards would be somewhere about the average. So you see, both sides were fairly steep. It must be thousands of years since these hills were moving, because only four to five feet down, the sand is congealed or turning to stone. Another peculiar feature is that about nine times out of ten there is a clump of trees growing in and about these breaks. This enabled me to see where the break was, sometimes up to three miles away.

The fact that there was a passage through the first one did not mean that you had a free passage up the next, for often we had to go back in the opposite direction for a mile or two before getting into a channel that would give us a run in our direction. At times this meant a great deal of walking, since it was no use moving ahead until a clear way through had been found. Quite often, I am sure, I walked more miles in a day than the truck covered. Once we found the distance between breaks, however, it nearly always ran true until that particular system altered, from ten to twenty miles perhaps. Then these sandhills would break up into a jumble of hillocks like the small patch of pyramid-shaped hills I described before, only they would stand about sixty feet higher. There was always a way round, but it took time to find it.

These sandhills were fascinating. Where did they come from and how? Why do they always travel east and west? There are no north-south sandhills. Why do they keep in straight lines and keep even-looking sides with the same V-shaped tops? They vary in colour according to the locality. In the main sandway, which is a thousand miles long and sixty miles wide, they are bright red. Other systems are a deeper red. Those where we were then are paler; sometimes they are reddish yellow. Here and there stunted gum-trees grow on them—the ghost gum. Its bark is very white and very rarely can one see two of these in sight of each other. This is why it gets the name of ghost gum, because it is always alone. In all my thousands of miles of travel in this part called No Man's Land (on the map) I have never seen a clump of them. I know a little about how to arrive at the age of a tree, and I would say that some of these trees are over three hundred years old; often alongside is an old stump of the parent plant. It gives me sufficient information to estimate the time since these sands were moving, for the life of two trees covers a period of from five to seven hundred years.

I have asked eminent geologists and others what they think is the cause of these sandhill breaks at such regular intervals. Most of them say that it is undoubtedly the action of the wind, but I think that answer is stupid, for if the wind makes a clean nicely-proportioned sand-ridge five miles long, then makes a break of two hundred yards and then goes on for another five miles, and continues on like that and sometimes for hundreds of miles, it is a clean-swept piece of ground hard in between. Just saying it is the action of the wind is a short cut to an end of a statement. In my opinion there never was sand in these breaks; they are always clean and that's why these sandhills are so interesting. They give food for thought, for they have a charm and mystery that grips one. I am always happy when amongst them; they fascinate me. They are not dreary wastes of desolation, as imagined by most people; they have rare flowering shrubs that compare in beauty with the most delicate orchids of the jungle.

10

Flowers of the Sandhills

OF THE MANY lovely flowering shrubs I would like to describe one. It is a beautiful thing, a brilliant orange colour with an edging which borders on a vivid scarlet. The cone-shaped flowers are from nine to fifteen inches long and about three or four inches across. The flowers are shaped like a huge honeysuckle flower and they are tightly clustered together on a cone, much as a banksia flower is packed together. I can tell you they are grand to look at.

There are also several kinds of hibiscus. Their foliage is soft and resembles the marshmallow and there are endless everlasting ones from a very beautiful bush, six to eighteen inches and twelve to twenty-four inches across. It has a pure white flower and is clustered together so tightly that it looks like a snowball. I have kept these flowers for months, and they have still retained their beauty. In the distance with a red-bricky background these flowers make a beautiful scene, so when I am toiling through the everlasting sandhills, you will know that there are compensations.

I have crossed this belt of sandhills three or four hundred miles further south after good rains. The bird life is simply marvellous. Only those who have seen it can believe it! Where these tens of thousands of birds come from is a mystery, for within two or three weeks of rain they arrive in great flocks.

It is almost impossible to believe, yet is a miraculous fact, that these bare, red-hot sandhills, shimmering in the haze, without a blade of grass or vegetation of any kind, after a thunderstorm of two or three inches can be transformed in two or three weeks into one of nature's most lovely gardens. Grass, buck-bush and countless everlastings spring up. The latter start to flower two weeks after the rain.

One cannot walk even a few yards without disturbing a nesting bird. Their mission in life seems to be to get as many hatchings as possible while the water remains. The most outstanding are the three different kinds of quail. They are common enough in other parts of Australia. They have always been classed as a game bird and they are generally scattered. They would usually be called plentiful if a hunter saw fifty in a day's walk; but here they were in countless thousands. No doubt lots of them were bred here, for the year I am talking of was the second good year running.

There was water in every direction you looked, although only small holes, ten to twenty feet across. These are called crab-holes, and they get their name from a crab-like creature that lives in this shallow hole. They are peculiar creatures, about as round as a two-shilling piece with two greeny-grey sides about the hardness of a soft gum-leaf. These sides expand when in motion like a concertina. The inside looks like a fish's gills, a pinky red colour and about a quarter of an inch thick. They grow to about three inches across. These crab-holes are teeming with the strange creatures. It occurs only when the waterholes last for a couple of years as they did on that occasion. There are years and years between good rains during which crab-holes are dry and parched, the fierce heat of many summers scorching and cracking them. But two weeks after rain these creatures come to life in their thousands.

And this of course is why some of the native birds come, like the beautiful spoon-billed white crane, of which one could see a hundred in two weeks' travel, yet in other parts of Australia it is almost an event to see a single pair.

There were other white cranes, the sharp, long-billed; and another kind, I only know them as red-heads and they are mostly in large flocks. It would be impossible to estimate their numbers. Then large flocks of golden plover abound. I had seen only four or five of these in my whole life. It is considered a very rare bird and yet they were there in thousands. There is every kind of water-fowl including black and white geese and speckled grey wanga geese.

The land birds also seemed to be just as numerous. These birds all come to the sandhills after the big rains. What puzzles me is, where do they come from and where do they go?

Then there were large flocks of big birds like the plain turkey and bustard, the tall blue crane and best of all, the brolga or native companion. I saw several flocks of these, from ten to fifteen in each. These stately birds hold a charm that no other bird possesses. They seem to get more out of life and always seem happy. The male birds are rarely quiet; they never seem to get tired of giving their trumpeting call, which is a very musical note and always in a treble key. They always seem to be looking for an excuse to have a dance. This dance, to my mind, is the loveliest thing that nature has produced in bird life. It is fine to see them line up in pairs, one male trumpeting as though calling the sort of dance. Then they all line up and join in the trumpeting: a series of treble notes; it is sweet music to listen to. Their light, quick, dainty steps and their spreading silvery wings as they dance and turn and swing and bow to each other are something worth going a long way to see.

There is one figure of this dance that I think is most charming and graceful. They line up in pairs, then one gives a trumpeting call and they step apart, spreading their wings. Then they bow their long necks and, still with their wings outstretched, they turn as though to fly; they then turn and take three funny little running short steps, then pause, then two more little short quick steps, then step and turning, swing right around with their spreading wings entwining and with trumpeting calls take two more long steps—like two long glides—ending by each bird swinging apart, turning right round as they separate; then come more trumpeting calls as they face each other bowing. They come together again, each taking a half-turn, then they line up as though they are going to fly.

This is very beautiful as it has such a lovely swing—the timing is perfect—it is grand—the poetry of motion of grace and ease. They will come through their sets of figures three or four times covering a distance of sixty to seventy yards. Sometimes they are all abreast in one line but their favourite way is to line up behind each other in pairs.

These birds are very tall, about three feet six inches to the top of their back. They have long necks, about eighteen inches to two feet, and their legs are about the same length. When standing they are about five or six feet tall. When their wings are at full spread they would be over ten feet across and the underpart of the wing has a wealth of most wonderful silvery grey lace-like feathers that are all ruffled out when dancing. It is indeed a grand sight to see a brolga's dance. I don't see any reason why the dance of the brolgas should not be adopted by human beings for it gives scope for grace of movement no other dance has.

I have been across the sand-ridges when they were devoid of almost any cover and in hot blistering heat and winds when one would have perished inside six hours without water, but even then they still hold their charm, for the vertical heat-rays going up for hundreds of feet cause fantastic scenes of mirage, an ever-changing moving picture, one of the most beautiful tricks of nature, reproducing scenes of placid waters and swaying grass or of what was or what is to be; reflections in the sky for man to read when he reaches a stage of development when his brain can grasp the meaning of it all. The people of those days will have lovely stories to read when science unravels the mysteries of the sandhills and their life. I always talk too much about the sandhills—it was dark when I got back to camp.

11

Lasseter Gets Surly

"YOU GO too far at the end of the day!" complained Sutherland when I got in. I replied that it was just as well for I had found a way through and we could get over two sandhills and all good going and be on our right course again.

This was great news; we were all anxious to get to Ilbilba. We still had plenty of water, but none of us had had a decent wash for days. We were being careful and saving a safety margin in case the Ilbilba water was not as good as we had been led to suppose. The direction and information had come to us secondhand from Bob Buck. Secondhand information is always doubtful and accuracy is not Bob Buck's strong point, so that was why the water regulations had been tightened ever since we left the timber belt.

Camp life had by now got into a routine. There were punctures to mend and we had a good fire. This night was calm as the night before had been and we all joined in a game of cards. Between deals I told them that I had come across several fresh native tracks. This information caused decided interest. One of the players started singing the ditty, "The Girl I Left behind Me".

Phil bit straight away and said, "They'll eat you first, you're fatter than I am." Phil was about as fat as a carving fork. Mickey did not miss any of this. When we were making up our beds for the night, he came to me and asked if he could sleep near the truck—he usually slept native blackfellow style with his own fire. I said, "You cannot have a fire if you sleep there, Mickey." "Me no wantee fire tonight, plenty warm alonga tarpaulin," he replied. So I said, "All right", for I know how afraid these station natives are of the myalls or wild natives.

I did not anticipate any trouble, although I did wonder whether Bob Buck and his tribe of half-caste children had been here; for if they had been in contact with the sandhill people they might have antagonized them as he has little regard for the native laws.

That was the only danger, and as I thought it remote, I took no precautions whatever. Phil and Harry, however, saw Mickey making up his bed and Phil asked if I really had seen fresh tracks. I replied that there was nothing to be alarmed over as from then on we should see them every day. Smoke-signals had been hanging in the sky all that day.

In the morning I was informed that every night Lasseter had locked the door of the truck in which he slept and which contained all our ammunition. That meant none of us could get to it in case of any emergency. I thought they were indulging in a little leg-pulling so took no notice. Later, however, I had occasion to go to the cab of the truck and sure enough he was locked in it. I gave him a dig about it but he said, "That's all right, you do as you like, but I'm not going to be caught napping."

We had a pretty fair run that day. I took a chance and passed a break, thinking that the next would do as the going was very good, but after about four miles we found the sandhills closing in, so we had to turn on good ground and go back to the opening we had passed.

We were in much better country now. There were patches of open grass and herbage, also some nice clumps of the snowball everlasting and other flowers. This was the first green grass we had seen; evidently there had been a fall of rain three or four weeks earlier. Although the sand looked as dry as all the other there was some dampness three or four inches down and this enabled the Thorny to get a footing almost anywhere.

The passages through the breaks were good excepting one where we had about five hundred yards of matting. There was just a touch of that soft silky sand in it, for it was always the front wheels that got bogged. By about 3.30 p.m. we had got down to only sandhills between us and the Ehrenberg Range—we felt sure that the camel-pad would be over close to the hills.

We stopped opposite a place that looked like the aerodrome and as far as our eyes could see there did not appear to be a break in this sand-ridge, so we decided to cross it. There was a slight depression where two ridges met. This would enable us to go over the ridge at an angle of about thirty degrees. We could have got over on the matting, but as this would be the main crossing for all time, we decided to make a track. We dug two trenches over the hill then filled them with spinifex grass. There was a good supply here; all of the ridges were covered with it. It was quite an easy thing to dig up with the long-handled shovel around the clumpy tussock. Just as much as one could carry of this made a fine bed. After we had finished the track in two hours, the truck went over in ten minutes. This made for a late camp. What had looked like the aerodrome was only a big, soft flat, well covered with grass. The old Thorny pulled up before it had gone many yards on it. I said, "This will do—campo."

Just then Fred drew alongside. He noticed smoke coming out from under the big truck on his side and yelled, "Fire!" Grabbing his long-handled shovel he started shovelling sand on to the seat of the fire. Of course the word "fire" made us all hop lively. We had three fire-extinguishers, two small ones and a very large one, but before we had got going, Fred had put out the burning grass and was down on his side pulling away any hot stuff that might light up again. We found that the heads of the grass had caught on one of the clamps of the hot exhaust pipe and ignited.

It was a quiet camp that night, for all were disappointed that we had not reached Ilbilba. This was the third day of the camel-pad and we covered only sixty miles so some idea can be got of how hard it was making and breaking a new track.

We were up early, and while breakfast was being prepared Fred went over the big flat and found the camel-pad again, which we had been forced off. Everyone got up tired. Even Fred's information about the pad did not stir us much.

Another small dispute occurred at breakfast. Harry, as usual, was last coming to his meal. In my hurry I had used his pint-pot to mix extra milk. He started grumbling and said, "Someone always takes my pint." I said, "I'm sorry, Harry old boy," and rinsed his pot out with hot tea and handed it to him. But he had a liver that morning and kept nagging at me. He was sore about the water being scarce and accused me of not being capable of handling it.

The atmosphere was a bit tense, and then George shouldered his way over to me, and said in his extra low voice, "Look out, say nothing, it's only liver." So I did not reply to Harry. George was sterling at these times. He always saw the other fellow's game miles ahead. The incident should have ended there but Errol promptly chipped in, and, in ten seconds, there were hot words flying. These two were always just on the edge of a blow-up;

neither of them was temperamentally fit for a trip like this. Coote completely lost his head, and said things that were childish and there was some hot talk of guns.

At this stage I stepped in and said, "That's enough", but Lasseter persisted and said, "That settles it. I am not going further. I will go back and reconstruct the party," and made some pointed remark about Coote's ability. At this Coote's Irish blood flared up properly. I raised my voice and said, "Here, cry off both of you. We are going to Ilbilba, so pack up; I am heartily ashamed of your conduct. We have broken our last forty-gallon drum of water, so on with the game." I turned to George and said, "Pack up, I'm off." Then turning to Lasseter I said, "Come on, we will find this water." I knew we must be very close as I had watched the flight of the topknot pigeons—they were here in great numbers.

In thirty minutes we could see the low bough sheds that the Mackay party had erected, and their big aerodrome. The sight of the drome acted like magic on Harry. He was all smiles and was lifting his feet again and all signs of his liver complaint had gone. It was the last three days that had got on his nerves, for we were expecting to see Ilbilba any part of each day. Although it was only three hundred miles by the map to Alice Springs, we had covered over four hundred miles of the hardest country I ever travelled.

Harry and I had a good look around the two big bough sheds. They were built with the branches of many kinds of trees and were still in good order. There was a windbreak covered with bushes for some sixty or seventy yards and this made the place look like home; nice and clean and room enough for everybody. What had been Mackay's dining-room would make a splendid place for stores.

As I stood on the edge of the 'drome I could not help but compare it with the 'drome we had made, for here was a clearing about a thousand yards long and six to eight hundred yards wide and not a stick or bush on it anywhere.

By this time the trucks were in sight and I directed Phil into a convenient place so that it would be handy to get to our stores-shed.

What a wonderful difference those bough sheds made! Just these windbreak fences and the bough sheds were the first touches of civilization we had seen for three weeks. It was hard to realize at first that such humble shacks could make such a difference! Harry was walking around laughing and talking with everyone. I reminded Errol of my order and asked him to respect it to the letter. It was easy to say anything this morning for we were all as happy as if we had found the "reef".

Phil said, "I suppose a fellow can have a bath now." I laughed and said, "Not yet, we will have to find the water first and see how big the supply is." Our directions had been that we would find the spring five or six hundred yards on the western side of the drome. Five minutes in that direction was sufficient to show that these directions were wrong, so Fred and I made for a gully in the hills. It was not many minutes before we found the camel-pad. We followed it for two miles and found a very nice little limestone hole that contained five or six hundred gallons of clean water. We just tasted it, but did not have a drink as there was a lot of animal and bird matter about.

It was grand to see the larger flocks of topknot pigeons. Their plumage is silver-grey, mottled with plentiful dashes of bronze, and with their pink topknots standing up three inches they looked very handsome.

We soon got back and told the party of the good news. I said we must wait until we got a load of water down before bathing, as I wanted to keep the Dashwood water for drinking. We unloaded Freddy's truck and, taking four empty drums, Fred and I went for a load of water, Mickey coming with us while the others unpacked the big truck. We got the truck as close as we could to the water-hole, and filled the drums. We also took the level of the water and timed the make-up. In thirty minutes it was up to the original level. This was good news, since this was to be our depot for a fair while and a constant supply of water was essential. By the time we arrived back, the big truck had been unloaded and what a mighty dump it looked!

It was hard to believe that all the dunnage could be stowed on the truck. No wonder we had had trouble in the soft places. There were over three tons on the ground and the weight of the truck itself was three tons. We had made a track with nearly seven tons, a track that would last for ever. It was a great achievement and the makers could have felt proud of the truck.

We had brought some wood down from the soak, as there was very little of it around the shed. A big fire, well away from the stores, was made and everything that would hold water was put by the fire to heat. Over two weeks of sweat and grime made us hurry to get the scales and stale smells off our bodies. We were like a lot of kids waiting to get our bath.

Phil said, "I've got a pretty dirty job on, will it run another bath?" I said, "As much as you like." So after we had all bathed and shaved and while lunch was being prepared, I took the splints off Phil's finger. It looked remarkably well and I put a stall on it, and Phil seemed comfortable.

Phil then informed me that he thought he had "big-end" trouble and would have to take the engine down for a thorough overhaul before we set out on our journey again. I told him he could have any of us to offside for him but he said that it was a one-man job, and if he wanted help he would call out.

Fred had so many things that wanted seeing to that he took advantage of this plant to effect repairs. It took them both a couple of days to repair the trucks. In the meantime the rest of us had been busy sorting out the loading. We had fixed the dining shed up with two tarpaulins to make it weatherproof. The store contained over two tons of foodstuffs and we would be able to leave over two hundred gallons of petrol in the dump. There had been a change in the cooking too, for the first thing I did when the trucks arrived was to get our yeast made. George's damper was getting on our nerves; damper is all right for a day or two but when it runs into weeks it gets very hard to take. Damper is made with baking powder that causes much belching—the common term out back is dynamite bread!

George was extra good at making damper. It was always palatable, when fresh. The first batch of yeast bread, however, didn't hit the ground for none of us had realized how much we had missed it until we tasted it.

We had over a hundred pounds of butter and for two days new bread and butter seemed the greatest luxury to be had. We also had a great variety of dried vegetables and we tried out the different kinds and boiled one of the six hams. It was a great treat to be on decent tucker, properly cooked. We were a very happy family in those days. Comradeship and team work were excellent and there was a tinge of excitement abroad for we were now within striking distance of Harry's wonder reef—"ten miles long, eight feet wide, over three ounces to the ton". It was easy to dream of untold riches, for if it were only half the size he claimed it would still be stupendous. One hundred million pounds worth of gold to the mile! Much fun was derived from the question—"What will you do with your millions?"

Fred lay back on his bunk kicking his heels in the air and started whistling that haunting tune, "The Girl I Left behind Me". Phil threw a boot at him. Fred enjoyed teasing Phil, who intended to marry when the trip was finished. George said, amidst the fun, "I suppose you will be chartering the best and biggest liner afloat to go home on?" "Aw, shut up," was all we could get out of him.

12

Golden Hornets

IT WAS at Ilbilba that we made our plans for the "Big Push". I wanted more speed, for we only had a bare month before the hot weather. Besides, Lasseter was anxious, for we knew that Michael Terry was out with his expedition on practically the same quest. We also knew of another expedition coming from Western Australia. They were all making for the same country, with exactly the same sort of maps, each one measured up carefully with tracings on it showing the locality of the Wonder Reef. Terry had camels and a motor truck; he had gone out from Horseshoe Bend station and would be from a hundred and fifty to two hundred miles further south of us. The West Australian crowd had only camels and, as this was the fourth dry year and they had six or seven hundred miles to come, we were doubtful if they would ever get through; they had started fully four months too late. Once summer hits the country nothing can travel on it; not even motor trucks. A ground sun temperature of 175 degrees will stop any engine. As Harry had the best "story" of the whole crowd, claiming to have been there twice and to have had correct bearings taken by a qualified surveyor, we had only to follow the information which was to lead us to the exact spot.

So far Harry had not complied with his contract to divulge this information to me. All I managed to get out of him was on the day we left Ilbilba, when I took him aside and asked him for the directions. He said, "Keep, first of all, a point or two north of west," showing me the position on the map of Lake Macdonald about 130 miles west of us.

Lasseter had been at his wireless plant ever since we reached Ilbilba and when Phil had finished his job he took a hand too. The receiving set had been working from the first day because the aerials of the Mackay party were still standing. Taylor and Lasseter worked on the complicated transmitter for hours but neither could make head or tail of it. The switchboard had over a hundred plugs, all wired, and if one were out of place the thing would not work. Harry had made a lot of notes, when in Sydney, of how each part acted, but I think his notes were too complicated for him to understand and, try as they would those two could not get the switchboard to correspond with the chart. Several times they thought they had it, but when linked up with our powerful battery it would not spark. Of course I had prepared the message they were to send, for I was more determined than ever to control what went over the air.

I had schooled Phil to learn the keyboard, so that he could check that my message was being correctly sent, for I had made it clear that only three of us would be with the machine when the message was being sent.

Later when George and I returned to Sydney, we submitted a copy of the chart to two wireless experts. Both agreed that it would take them several days before they could put it together, and they were of the opinion that the system was wrong and were doubtful if it had ever worked. The firm's excuse was that it was an "experiment", and they had informed the company all along to that effect, but all the same, they got over £100 for it.

On the fourth or fifth morning of our arrival at Ilbilba, Coote and Colson departed, Colson for Alice Springs and Coote to pick up the plane at Taylor's Creek about a hundred miles back. George, Mickey and I had gone ahead to clear a new track, for the track we had made coming in was out of line, besides being all soft going. We had previously decided to make this new track direct to the crossing of the sandhills.

By the time Fred had caught us up we had it nearly done, not that there was much to do. Most of the work was throwing many dead trees out of the way; it was good hard clay country. With a final handshake, I gave Fred instructions that he was to go through to the Alice with the mail and bring another load of petrol through; also he was to pick up eighty gallons of water at the Dashwood, just in case the Ilbilba water would not keep in cans. It often happens that such water will turn bad if carried for a week or so. I told him that when he got back to Ilbilba he was to bring water and petrol out to us. I intended to give him Mickey as an offsider when he got back and would have liked to have done so, but we both knew that once Mickey got within a few miles of Archie's place nothing on earth would stop him from leaving our party.

The arrangement that I made with Coote was that he was to take no chances about getting off the 'drome, even if it meant waiting the three full days. I did not like the small patch we had cleared. It was surrounded with traps. I had not forgotten the previous attempt to take off which at first had looked like certain disaster. Taylor had told me that the way Errol had pulled the plane over the tree-tops was the wildest sort of gamble, and a thing that comes off only once in a dozen times.

As it turned out the arrangements I made with Coote were the maddest arrangements ever made. We had agreed that we were not to bother about him for three days; if he did not turn up by then we would know that something had gone wrong. It seemed all right when we went over each detail several times, for he always was covered by the fact that Colson would be with him. I shall never forget those three days of waiting; it was a torment.

I realized now that the agreement to wait the three days was a bad mistake. If the plane was forced down between Taylor's 'drome and Ilbilba, the pilot, even if uninjured, would be in a very bad way. Two days, even in winter, without water is almost impossible. The weather was perfect, except for flying, as it was too calm. I had misgivings that things were not well and the three days I put in waiting were the most trying experience I ever had.

I could not leave the camp—reading was impossible—I was too uneasy. I just walked about listening and listening. A dozen times the big, black-striped golden hornets fooled me, for they make a drone very much like a plane in the distance.

These golden hornets are most remarkable insects. Their bodies are about two and a half inches long and their wing span about three to four inches. Their black and gold stripes make a charming colour study as they form rings round the body. They are stately looking, with long legs—over two inches—and are incredibly quick in their movements. They have extra keen sight and hearing. Their gameness would be hard to equal, for they are always ready to fight anything, no matter what its size may be. All dogs know by instinct these hornets are their deadly enemies, and whenever they can, they take a lot of trouble to keep clear of them.

In fixed mining camps I have often succeeded in taming them. This is, of course, rather ticklish work and some care and much patience is required to succeed. My method was first to catch a nice fat grub and stick this on the end of a reed. I would get an insect that had settled and very gently and slowly work the bait towards it. Once the hornet sights the insect, it is amazing to watch his caution and amusing to see how he will run around many times, inspecting it from every angle. Then he turns around and scratches sand and dust with his powerful hind legs and kicks it all over the bait and then does some more skirmishing round. The slightest movement on my part would cause him to fly around me, inspecting me from every angle and ready to attack at another move. This is what makes the operation so delicate; a hornet's sting is twice as strong as a bee's. But once you get the hornet to take your bait the rest is only a matter of time. When out working in the day we used to save everything that looked like a feed for the hornet and put it alive in a tin till we got home.

They seem to be able to think, too, and count as well, for I have often tried out little experiments to see if they could remember. I would bring home three insects of a kind and show them in the tin to the hornet. He would fly away with one and then I would remove the others from the tin and what a fuss there would be when he returned for the next load. He would go round and round the tin, scratching the dirt and searching under anything that might conceal the insects, indicating that he could remember there was more than one. He would also fly away with one, two and three and would not return looking for a fourth unless there was still another in the tin.

These hornets are as good as watchdogs in a camp; they attack any stranger entering, though they leave the owners alone if they are carefully treated. On mining fields, which are also on golden hornet country, the miners are always careful to search for hornets' nests before chopping into the trees to make camp. One can, of course, light a fire under the chosen tree and get rid of all the hornets, but most miners prefer to go to another tree rather than destroy them, because they are not only rare but are quite an interest in the camp and provide a bit of fun.

Their nest is a neat affair, about two and a half inches in diameter and built of mud. They place a few live insects or grubs or spiders in their nest and lay their egg on top of the insects. Then they cover the egg with more live insects that seem to be paralysed; and, when hatched, the young hornet lives on these insects until it has the strength to break the seal that covers the end of the nest. When it emerges it is ready to fend for itself and does not need attention from its parents.

13

Anxious Days

OUR ANXIETY increased as the days wore on. At night I would sit for hours, for our arranged times to send signals were 7.30 p.m., 8 p.m. and 9 p.m. At eight o'clock I burnt a twenty-minutes flare, then ten minutes white, then ten minutes red light. These flares were powerful, 32,000 candle-power. I signalled from a high point above the camp that stood eight hundred feet above the plains, and stayed up there until 9.30 watching for the faintest sign. Phil would flash his spotlight into the sky. It was almost as powerful as a searchlight for he could shine it on me while I was sitting on the highest rock. George walked away a couple of miles to see what effect it had in the distance, and he said that the top of the hill stood out in bold relief and estimated that it could be seen as far away as the hill was in view.

It was on the second night while I was on the hill that Lasseter thought he had got the wireless working. He did nothing but rap out our calling signal then send the S.O.S., and kept on repeating. Phil had the earphones on, listening in case our calls were getting through. He had nearly three hours of this.

When I returned Lasseter said, "I think I can make the thing work now." I was surprised, so said, "Give her a buzz and show me." But Phil chipped in and said, "It's a good thing it's not working for the cow has done nothing since you left but give our calling signal and then rap out the S.O.S."

Harry said, "Well, you never know, Coote may be in need of help."

I explained, "Don't you understand that I am the only one to say what is to be put over the air?"

He tried to be funny, saying in a little voice, "Did you think I would run on top of the hill and ask you?" So I got Phil to work the calling numbers while I connected the wire with the aerial to make sure nothing had gone out. I would have died with shame had the S.O.S. gone out for I was bound to keep the agreement made with Coote; and it was quite probable that he was waiting for some wind to take off. I was very relieved when Phil declared the radio "dead".

I then said, "The machine goes back in its box tomorrow, make no mistake about that." But next morning Lasseter did not want to pack it away and started arguing so I said, "Very well," and got the axe and stood within striking distance of it—"Do you pack it or not?"

He said, "What do you intend to do if I don't?" I lifted the axe and said, "I'm putting this into it, then I will be sure that you won't be monkeying with it while I'm away." He could see I was not bluffing and said, "Oh—all right." It was not unpacked again until it had been sold in Alice Springs—for a tenner.

I had gone into the wireless question with a great friend of mine, a member of our mining company, and he had solemnly warned me to be very careful what was sent over the air. He said, "You never know what stunts may be worked with the share market, and, if you don't find the reef (and there are a lot of people here with dearly bought shares), you are sure to be made the scapegoat." I knew what he meant. Once I found that Harry did not intend to honour our agreement I shut the wireless up. I found out afterwards that the mail that went back when Captain Houston left us contained information concerning the reef and the expedition which had not been submitted to me. In our agreement every member sending mail away had agreed to show everything to me; but Lasseter had sent word to Sydney that he had seen country that he recognized. The company accepted that information, knowing that it was not OK'd by me.

I have told you how I was attacked by a parliamentary member. The same man would have given a good price for information that a Minister's brother was working a "stunt". The Government's worries and responsibilities were hard enough then without any mining scandal. The eyes of the whole of Australia were centred on the expedition, and England was keenly interested because all knew that if this reef were found, Australia would be flooded with new people, and everyone was praying in this dark hour of depression that it would be found. The thing had become a "national straw" that everyone was desperately ready to grasp. Stark ruin was being faced by thousands, and every little bit of information about the expedition was eagerly sought. I tell you this to show how serious my responsibilities were, and that's why I stood my ground so firmly about the wireless set.

On the second night of waiting all hands showed the high tension of the worrying, waiting and suspense. Phil was untiring. He kept the searchlight playing over the high tops from dark until 11 p.m. I am sure it lit up those hills so that they could have been seen for fifty miles. What the natives thought of the flickering light would have been interesting to know, for it looked very weird even to us.

On the third day I was nearly mad. I cursed and cursed those arrangements I had made with Coote. In the afternoon we stripped the Thornycroft and got it ready, for as soon as darkness came the waiting period ended. We carefully selected what we thought necessary on such a journey and took the truck over the sandhill and waited until the sun had set. Then I was free to act and we started on the wildest and roughest trip ever made by a motor vehicle.

We left Harry and Mick at the camp, just in case Coote should arrive while we were away. We soon realized what a big job we had taken on, for with only two men handling the four mats, it was cruel. The darkness behind the truck was intensified by the glare of the bright headlights and spotlights. We intended to go right through that night if possible.

We made the discovery that the truck was harder to handle without a load. The four driving wheels would get a jumping, bouncing movement on and would dig themselves into the sand. We had plenty of bags aboard so we filled a dozen with sand weighing in all a ton; these we put over the driving wheels. This steadied the truck and gave the driving wheels a good grip. We found, too, that the fixed headlights were not much use in taking the many sharp turns in that twisting road. So I took the powerful spotlight in my hand and searched ahead for the bends and, while Phil was pulling the truck around, I could manipulate the spotlight so that he could practically see around the corner. That made it easier.

After that we had very little matting to do, for the old girl would fairly fly over the sandy patches. It was a great relief to George and me; I am sure that otherwise we would have knocked up before midnight. The road was surprisingly good; we sometimes touched twenty miles an hour, but when we hit the timber belt, the fun started in real earnest.

The road through the scrub was all shapes. When we were coming out we had dodged every tree we possibly could. That was all right going slowly but tonight, when we were able to go faster, the winding track made all the difference. Although the lights were powerful we could not always see the short bends and often crashed straight through.

It was a thrilling ride, and afterwards we all wondered why one of us was not staked! Once George nearly got his issue. A stout, sharp stake drove right through the opening on his side and stuck through the boards at the back of the cabin. All he said was, "Spare me days, Phil." Before he could say any more there was another crash, this time on my side. There had been just one panel of glass still intact but this crash shattered it. The glass flew all over me. "Some hit," Phil said. "Are you hurt?" I put my hand up to stop the truck, afraid to answer, for my shirt was open and I could feel a lot of broken glass inside. He pulled up and took the spotlight from me and, very carefully, I undid my belt and pulled the shirt out. There were no cuts, just a few scratches and some minor cuts about the forehead and hands. Phil, too, had a cut, an inch long, on his hand but we soon got going again. Phil asked whether he should slow down a bit, but George and I thought if we went slowly it would mean cutting down several of the trees and that was no job to tackle at night so we both said, "Let her go."

Phil was most concerned about the spotlight. "Mind you don't break that," he implored us; "the headlights may go at any time!" He was driving now, straining every nerve, and as we watched the sweat break out we marvelled at his skill and dexterity as he twisted and turned that wheel through all the contortions of the almost unseen track. George tried to express his admiration and, every time Phil made a good "save" gave him credit by shouting, "That was splendid—great work, Phil!" but when George started singing, "Off to Bonny Scotland", Phil fairly roared: he yelled at him, "Cut that out, how can I concentrate on this road if you yell that stuff?"

We were hitting some trees fairly hard now and we had to clear one tree off the bumpers. Phil thought that if he drove a bit faster he would throw these trees. We consented but actually we were both scared stiff. About eleven o'clock we hit a bunch of trees at about twenty miles an hour. I can tell you the shock was terrible for it stopped the old truck dead. Both headlamps were smashed, and I had dropped the precious spotlight and all the glass in front of Phil had gone! The three of us were dazed for a second or two and were afraid to move or speak. The spotlight had fallen on its face so the darkness was like ink.

I said, "That's the end of the section," and lit a match. Then we were all talking at the same time. I noticed with relief that the spotlight was still alight so thankfully I got out and picked it up. Phil and George stayed where they were. Many small bits of glass were caught in their shirt-necks. There was some blood but only from minor cuts. The marvellous part of it was that when the tree hit the glass, it threw it up high and nearly all of it stayed on top of the bunk.

The three of us were pretty excited, I could hardly hold the lamp I was trembling so. I had missed all the shower of glass but had knocked my head pretty hard. They bent down and shook the glass out of their shirt-necks and then we had a look round. A pile of trees were stacked up on George's side and the top corner of the cab had been smashed in. It looked pretty bad to me as we stood there.

Phil decided to see if he could back out. It was quite easy. We soon had the trees out of the way, and Phil examined the headlights. Only the glass and bulb had gone so he stuck another globe in. George and I examined the cab. Things were better than they appeared at first. It was shaken a good bit, but we thought that it would stand.

When we moved off again we decided to be particularly careful as there was no protection now at all. George and I could move a bit if we saw anything coming, but Phil was a fixture. We ran on for about half an hour and then stopped at one of our old camping places and decided that this going was too risky. We had covered about ninety miles. None of us had quite recovered from the big crash. We discussed broaching the brandy in the first-aid chest but did not get any further than that. We were thoroughly exhausted, so made some hot tea and turned in. It was a long time before we got any sleep; in fact, they were only cat-naps that we did get, but it was remarkable how fresh we were when we got up, although we looked three very wild fellows for we had all forgotten to have a wash.

After we had washed in hot water we found that our cuts were only scratches. As soon as it got light enough we cut a tree down and trimmed up a length. Phil bolted it on to the side of the truck and end of the cabin; this brace held George's side a bit steady. Before sun-up we were away, but had not gone very far when we decided we would have to close in the front with boards, only leaving a small space for Phil.

It was worse sitting behind the board, for one was always expecting a crash; but the starch had been knocked out of Phil, and he did not attempt to hit any more big trees. We had a bit of bad luck about nine o'clock and punctured four out of the six wheels. So while Phil and I were fixing them, George boiled the billy and we had a decent feed; we had not eaten much breakfast. We got going again about 10.30 and did not pull up until we got to Dashwood Creek, arriving there about two o'clock. As we ran into the east-west runway we could see the track of the plane where it had run down in the take-off.

Phil got off to look at the tracks. Our hearts sank into our boots, and I cursed again the maddest arrangement ever made and thought, "Four days! There is very little hope for him."

When the truck got down about two hundred yards we saw the plane. What a relief! It was, however, standing on its nose and upside down and forty or fifty yards off the runway, tipped up on the bank of the big creek and lying on top of a big old rabbit warren that was all big holes where the natives had been digging them out. Both wings were crumpled up. We made a hasty survey. Phil asked me not to touch anything as he would have to make a report to the Department. He examined the cockpit and decided that Errol might have had a bad smash, for the front cockpit was very badly crushed in. We could see the track of the dried blood and Freddy's footprints where he had carried him over to his truck. That gave us great relief, for we knew Fred would go with all speed to the Alice, and wondered how badly Errol was hurt. There was no note on the plane but I walked over to where Fred had dumped three crates of nine-gallon water-drums and saw a note tacked on the crates.

It told briefly what had happened: that Coote was not badly hurt but had a deep cut in the fleshy part of his thigh. Phil remarked that he had got off lightly, considering the position in which we found the plane.

There it was, lying on its back, the undercarriage still intact. What a helpless thing it looked; it seemed so flimsy! We wondered where it got its strength to withstand the pressure of the wind. There lay our pretty, painted *Golden Quest*, a crumpled heap, and now that I knew the pilot had not suffered much damage I felt glad. This was the end of one of my biggest torments, as I never had any faith in it. I had been stuffed with all sorts of tales of its ability. It had cost a thousand pounds, the mileage flown had cost the company about £3. 10s. per mile and had delayed us about ten days, which cost another £150.

I feel sure that when the air conditions of this type of country are investigated, it will be found that it is a very treacherous place for light planes, especially in hot weather when the heat waves are wriggling upwards. Later I discussed this with Charles Ulm who agreed that flying over the hot sand was a tricky job and said that he had experienced it when he struck one of the hot air channels which made him lose up to 2,000 feet elevation. I have no doubt that Coote tried to take off in similar conditions.

That evening I wrote a report of the crash to send back to the company. By the note Freddy had left on the crate we could expect him back any time now. Phil had decided to dismantle the plane and we would pack it on Fred's truck and send it back with him to the Alice. So, with a mail on hand, we were all busy for a couple of hours.

I begged the Sydney crowd not to send another plane. If they wanted to help us half a dozen fast-riding camels would get better results. But only extracts of my report were read at the meetings and when I returned to Sydney I heard to my surprise people saying that it was I who had recommended that another plane be sent out as quickly as possible.

By next evening we had dismantled the plane and, with some repair work done, the Thorny was ready once more for the track. We got a decent feed ready, and while it was cooking we fixed the big tarpaulin over the end of the truck and made a canopy under which we rigged our bunks, for the night was chilly.

That evening was spent listening to Phil's tales of wrecks he had either seen or heard of. We were all hoping that Fred would turn up tomorrow. Taylor's Creek was a bad camp. We should have called it Disaster Creek, for here the truck broke down when we lost the captain, and here again we were waiting.

14

Fred Colson Returns

EARLY the next morning I heard the faint drone of a motor engine; so I got up and made the fire and put the billy on. Phil sat up and said, "What's biting you?" I said, "I heard the drone of Freddy's truck a while ago." He said, "Get back to bed, you're dreaming, you've got buzzing noises in your head. If you're cold, use the small tarpaulin—waking a fellow up in the middle of the night!"

I listened, but could not hear the drone again so after a while I put the billy to one side, stuck a couple of logs on the fire and went back to bed. Later on it was George's turn. He was yelling, "Come on, get up, here's Fred." We listened again but could not hear anything, so George came in for a lot of abuse too. Phil told him that we were both balmy and had "motor-engine noises on the brain". However, George put the billies on again and said, "I heard Fred an hour ago." Then we got it! Phil said, "Oh yes, have a yarn about it. You are as bad as two people swapping dreams. Anyway, what's the time?"

George said, "Twenty to five," and Phil said, "Oh hell, get back into bed and stop talking." So George stood by the fire for a while then said, "That's funny, I haven't heard it since I got up."

"No, you long Scotch cow, you won't while you are awake. Get back to bed and have another dream about it," growled Phil.

So George came back to his bunk and just threw the covers over himself.

About half an hour later we heard it again. George bounced out and said, "Did you hear it that time?"

That time we all did, so we got busy preparing breakfast. Fred had driven right through. He knew we would be waiting for him. But when breakfast was ready, he still had not arrived and it was after six before he came in.

When I first heard him he must have been twenty-five or thirty miles away, and about fifteen miles when George heard him, for he had not stopped for four or five hours. While he was having his breakfast he told us the story of the accident to the plane. He said that there was practically no wind for Coote to take off with and what little there was was coming from the east, so they decided to lengthen the runway another seventy yards on the top end.

"After running the engine for a few minutes," he went on, "we shook hands and I said, 'Be a good boy—there is plenty of daylight.' So away the plane went, but it was three hundred yards before it lifted. That only left him eighty yards to get up clear of the trees which were from fifteen to twenty-five feet high. The engine seemed to change tune. Then it wobbled over to the right side, hit the tops of the trees that slewed it round to the left side and it hit the big limb on that tree" (pointing to it) "then it turned over. I ran as hard as I could and got him clear. I was afraid of fire—there seemed to be smoke coming from the engine. Coote was doubled up and I thought he was done for. He looked pretty bad and I lost no time rigging a bunk in the truck, making it as comfortable as I could with leaves and boughs. Then I scribbled a bit of a note, stuck it on the crates, and had him on the way to Alice Springs in twenty minutes. He was bleeding pretty freely from a cut in the thigh."

Fred said it was a hell of a drive and every time he stopped he could see the blood-patch getting bigger through the rough bandage he had put on.

"Coming towards morning," he went on, "I could hardly keep my eyes open so I sang all the songs I knew to keep myself awake, and in twenty-two hours I had him in hospital."

Fred handed me the mail. In it was a letter from Coote asking me to send to Sydney for another plane. There were also quite a lot of instructions from Sydney.

I had to laugh for one part said, "As soon as you receive this, act at once, then send a wire giving me results." These instructions had to be delivered by motor and we were over three hundred miles out. It might be a week or a month before a return-mail went in. There were lengthy lettergrams with all the bunk in the world in them, for, according to the Sydney crowd, half Australia was rushing to "peg out" their claims before we got there.

I had already written my report, and did not feel inclined to alter it. I just put a postscript stating our plans and that Coote had asked me to send urgently for another plane. I answered Coote's letter and told him we did not want any planes, but to buy a couple of pairs of heavy walking boots and to see if he could hire five or six fast-riding camels, also a native who knew poison bushes and was an expert with camels. I said to get bullock camels if possible, no cows at any price, and then he could join up with us and be of some real value.

I was always sorry that I did not take six good light camels with us. With a camel I could have prospected near and far and at the same time kept a look-out for the best track for the truck and could always have been ahead of it.

Fred said, "I had a bit of an accident coming out—lost the nut off one of the back wheels. The wheel came off but I managed to rivet it on somehow and I have been expecting it to come off again. It happened about midnight, when I was in the little hills. I had to unload the ton of petrol, at least most of it unloaded itself. I dug out a hole and backed the truck into it. That's how I managed to load up again. See what you can do with it, Phil." He went down the creek where it was quiet to have a bit of a sleep. Phil found any amount of work on his truck, and three hours after, when Fred returned, he was still finding little things to fix, for this kind of work is hard on motors.

After lunch, we loaded the remainder of the plane on Fred's truck and transferred the petrol on to ours. That night we made plans, how and when Colson would find us, and then drew up a chart showing the things we would want from the base. We worked out also the time we could expect him back, giving him a margin of three or four days, and told him it would be a long time before he could return to the Springs, so he had better make his arrangements accordingly and bring along half a ton of water from Houston Soak at the Dashwood.

The talk drifted around to Coote, and the company's extraordinary letters. We discussed the possibilities of any expedition "getting there" first; but Fred and I were of the opinion that none of them would dare to go out into the country we were bound for. This was the fourth dry year and those coming from the south would have a two hundred mile dry stretch to cover. The crowd getting out from the west had no chance at all so we came to the conclusion that

Coote and the Sydney crowd were getting all hot and bothered about something they knew nothing about. Then I gave Fred the mail, instructing him to send the wire and post the letters and, if the Flying Boy asked questions, to say he didn't know anything.

15

We Find Our First Natives

NEXT morning Fred got away just on peep of daylight. We saw him safely over the creek. His load looked a queer one, with about ten feet of the plane hanging over the end of the truck. Then we started back to Ilbilba.

For the third time we camped again in the centre of the timber belt. As this was a very dry stretch we left a nine-gallon drum of water and two empty fruit-tins, I think we left tea and sugar also. We put down water like this in many places, for with motors to deal with we never knew when we might be on foot!

At daylight we hit the track, but did not make very good time through the thirty miles of timber—we had lots of stops in order to clear the wreckage from that wild night-ride. We pulled up and had a look at the place where we had the big smack, and wondered that we had not done more damage; three or four of those trees were pretty hefty. We had pleasant recollections of it too; I think that's where we had our longest laugh.

Now that the truck had travelled this track three times it looked something like a road. Just after we passed our old camp at Mt Liebig we saw seven or eight of the wild horses that run on this country. The stallion valiantly stood his ground, stamping his feet, quivering all over, defying us while his little band of mares made for the thick timber. Considering that these horses had been inbred for about forty years they were remarkably fine specimens. Of course we did not get close enough to view them clearly, but as they galloped away, they looked nice—especially one filly, a really bonny animal.

A remarkable thing is that the natives here will not eat horseflesh although they ate all the cattle that were brought out some twenty years ago when there was an attempt to make a cattle station. These natives, incidentally, cannot be recruited for labour. Their country has an abundance of food, and besides being independent, they are different from the tribes elsewhere in Australia. I like to call them my people of the sandhills. They are intelligent and have a religion which is recorded on remarkably long sticks with a lot of writing on them. These sticks are about twelve feet long, an inch thick, and six to nine inches wide. They are kept in caves and are very sacred, having the history of the tribe on them and are only used when they worship.

The fact that they will not eat horse or camel flesh is because it is not of the cloven hoof, this seems to be part of their religion and connected with ancient history. Their women do not take any part in their religious rites. The men go outside their own boundaries to get their wives, and their own girls, when they are eligible, leave the tribe to marry.

Very little is known of the tribes that live outside the boundaries; they appear to be outcasts living on the edge of the boundaries. These are the people who are first encountered by the visitor to the region, and unless one knows what they are one gets quite a wrong impression, for the better-bred men are shy and hard to contact. Rip Van Winkle, of whom you will read shortly, and his band were of these outcasts.

We had a remarkable run back to Ilbilba. We did not use the matting once, though we got a few punctures, and we arrived there just before dark.

Lasseter received the news quietly and, after tea, he was told the full story. When I told him of my report to Sydney asking them not to send any more planes, he was pleased, but said that, with another man a plane would be all right but that Coote was impossible and he hoped he would not return with Colson.

Next day was a busy one renovating the truck. Phil and George knocked our best case to pieces. They removed all the splintered board and put new timber in. They also took out most of the board from the front of the truck. I was pleased at this, for it was deadly sitting behind these boards. Phil removed the mudguards, too;

they prevented him from seeing the front wheels on the matting.

Harry, Mickey and I put up a staging to store the water and petrol. It was built six inches higher than the truck to enable Fred to load the forty-gallon drums of water and petrol easily, as he would not have an offsider. We also made a track with grass and bush so it would be easy to drive close up and start off, for Ilbilba was a very soft sandy patch. Then we all got to work in the store-room packing up cases and labelling them so Fred could see at a glance what he had to bring along.

Next morning we took the big truck up to the waterhole, also all our dirty clothes, and we all had a final clean-up, but the soak only made ten gallons per hour so we had to wait until the following morning before we got our load of three hundred gallons. We placed about half of it on the staging, and there were no mistakes this time of not taking enough water.

We soon packed the truck ready for the big push-out, had an early lunch, and were ready to start when we saw a native coming from the direction of the soak. The only costume that he wore was an old, greasy felt hat. He came empty-handed and walked in as though he was part of the show. I thought, when I saw him with that hat on, that here was a chance of finding out something of the country we were making for. I felt sure he would be able to understand and talk a little. I exhausted all my sign language on him, also different kinds of ground signs, repeatedly giving him the water-sign; he evidently did not understand. I was not altogether surprised, because what I was saying by sign belonged to a tribe further south. His reply was "Ningie, ningie!" which puzzled me but I learned later it meant that he wanted tucker and matches. He then made a lot of dog-tracks on the ground and pointed south to show that was where the dingoes were—he took us for "doggers". He was the poorest specimen of native that I have ever seen, quite evidently an outcast from some tribe. I thought him a bit touched in his upper storey and still think so. George Sutherland later christened him "Rip Van Winkle". He was dirty and skinny, and while he was talking to us his eyes were covered with flies. It was our first glimpse of the natives around there and we were certainly not impressed with this one. If he were a true specimen of the natives in these parts (later we found he wasn't), we were likely to strike trouble for it was quite plain that this one did not want us to go west.

I was sure he understood my water-signs and it was plain enough to me that to take or interfere with water further west was likely to cause ill feeling. Mickey, who was an Alice Springs native, could not understand him any better than I could, although he showed that he was afraid of him. I noticed that Mickey was many shades darker in colour than this Ilbilba fellow.

I gave up trying to understand him, regarding it as wasted time, and we gave him some food, a lump of one of Harry's old dampers. I reckoned it would give his digestive organs some work and threw in a ham bone with about three pounds of dried meat on it. So off we started, and had only got two hundred yards when out came the matting. For half a mile we had hard work. This country had a tough crust on the surface, but was like flour underneath.

We got clear of it at last, and with relief found a good channel. This lasted for a mile or two, then crossed over into another channel which meant more matting for another half hour.

Next we got into a patch of timber. The going was good but, as before, clearing a way through was slow. Although we did not have to cut many trees, there was a lot of timber about that we had to drag out of the way. I could see this timber belt extending up for three miles or more, so kept on the outskirts of it, then sand showed up ahead. I went up to it and saw at once there was no hope of getting further. I think George was with me. I said to him, "It's no use going along blind like this. I will have to go on foot and find out what's ahead."

About forty minutes later I returned and we had to take the truck back three or four miles and then cut another track through the timber.

While they were doing this I went on to see where the next break occurred, but in a mile I came onto spinifex and found some hard going with small hills ahead, so went back to the end of the sandhills and met the party just as they got the truck through the timber. The queer-looking native had followed us and when we got the truck around a sandhill we saw him with a woman and three children. Two were crippled, one being quite badly deformed. The other one, a girl, seemed all right but stared at us like an idiot. The mother seemed a fine stamp and was very young. These people had made some previous contact with white people—I suppose the Mackay party. Anyway, this woman could not speak but she soon conveyed to us that she wanted matches. She made a little mound of thin sticks then went through the motions of lighting it and the old man with the hat on kept on whining "Ningie". I tried to talk to the woman but she was too shy or maybe afraid to talk to any of us. This was rather an item of interest for the rest of the party as it was the first time some of them had seen a naked native woman.

Standing there with the innocence of a child, although a mother of three, she had a remarkable figure; her breasts were not hanging down as I have seen in other mothers of her kind; her bosom was nice and full and firm. One would not have thought it had seen so much service. She had rather a shapely head. Had her hair been clean and well brushed I think it would have been a dark brown. She had lovely big brown eyes covered with very long, even lashes, nice high forehead and remarkably well-shaped eyebrows. Her nose was broad though not at all flat, with extra large nostrils. Her teeth were just the most perfect things imaginable, not heavy-lipped; in fact the upper lip had a nice curve and the mouth was not big. Considering the woman was humping these two monstrosities about, she looked remarkably well, and I thought she was no more than eighteen or twenty years of age. She was just a bit hippy with a well-developed sit-down that only added charm to the perfect curve of her back and figure. But dirty! Goodness me! She was scaly and hummed a treat, but in spite of it all she looked nice, and it was beyond me to know how a rotter like her husband ever got hold of such a lovely girl. She had a wonderfully winning sort of smile.

I called to one of the boys, "What do you think of her?" He just held his nose and cried, "All aboard!", so we left them and ran on to the hills I had seen.

I had set a batch of bread that morning so made an early camp, 4.30; we had come eleven miles as the crow flies and we all lost some sweat over it. Today had been our best team work; everyone seemed to buck in and know what to do next.

We had an open camp, clear for hundreds of yards. Wood was a bit scarce but with everyone helping, the fire was soon going. George looked after it, while I knocked up the batch of bread. This was the first time we had set bread when travelling and it had risen nicely. About ten o'clock it was cooked and we had enough for three days.

Just before sunset Rip Van Winkle came up to the truck again. He came the match stunt again—natives love matches. I told them not to give him any more tucker, but someone threw him another box of matches, so off he went. I kept an eye on his direction and as soon as he got to some rocks and bushes he took cover and, in a little opening to the right about two hundred yards from where he took cover I saw him pass bending down low. This was what I wanted to see. Had he walked away upright, I would have thought there were more of his kind about. I then watched a rocky point on the end of the hills, and saw just a faint smoke—so that was all right, there were only Rip and his family. They were evidently afraid of other natives seeing them and I saw by that that he was "out of bounds"—meaning that he was off the country allotted to him. It is native law only to hunt on one's own ground. I think that law applies all over Australia. If he had been on his own country he would not have taken cover like that. This information was handy, for it helped define the boundaries of the country.

16

Lasseter's Diary

IT WAS a happy camp that night; we were all pleased to be getting on with the job. Harry was particularly friendly. He invited me into the cabin of the truck—this was his bedroom. The electric light was burning, as he always did his writing here. Every night since we started he had written in his diary, a nice-looking volume about eight inches long, six inches wide and perhaps two inches thick. It was well bound, with a very fine black morocco cover and exceptionally good paper. He said, "I have been doing some figuring while you were away, and by my calculations we are approaching my country. According to the bearings taken by Harding when we were out here together, the reef is somewhere about a hundred miles ahead. I am hoping to be able to recognize something about this lake for, in my recollections, there was a salt lake handy to where we took our bearings."

I said, "Were you on the reef when you took your bearings?"

"No," he replied, "we were some distance away. Where we took our bearings there was plenty of water. We had been hard pushed for water when on the reef country, so were not able to stay and only just examined a portion of it and took some samples, then had to make all speed back to the water."

I then asked him how far and in what direction was the reef from the water. He was vague and admitted that he was not certain, but they were away from the water about three days and he felt pretty sure that they got back to the water in one day. Then he got out the map and showed me the position of Lake Macdonald, and together we examined it, using a compass and rule and we made it out to be 110 miles away.

I then said, "Where did you take the reading from?"

"From a very high hill," he answered.

I then drew his attention to Mount Marjorie marked on the map. It appeared to be about thirty miles north of Lake Macdonald, and he said that that might be it.

I felt greatly relieved, for this was the first definite statement he had made since we started out.

I did remark, "Well, Harry, that is something definite."

He answered, "Yes, this is the first time that the party representing the company has been alone."

I then asked him to give me a reading of the bearing but he refused to do this, for he thought that he had told me enough and said that I would have to trust him for a little while longer. He explained then what he had said in Sydney, that when they took their bearings the watches were out by an hour and fifteen minutes, and they found this when they arrived at Carnarvon. As they did not know how long that took they were not able to use the correction.

I don't think myself that Lasseter could take a shot at the sun for, at Ilbilba, I had asked him to do so when we had plenty of time. In the morning when we arrived at the bough sheds I spotted a notice nailed on one of the uprights giving the correct position. Evidently one of the Mackays' party had thought it might be useful to someone. This I pulled off and hid away, for it was something by which I could check Harry's knowledge and I would know, for certain, whether he could use the sextant correctly. However, the little blighter refused to take a reading, saying that it would be of no use to us. I have often wondered since if Harry saw this notice first. If he was uncertain of himself and carrying on a big bluff, naturally he would refuse to show his hand.

This night, however, Harry was talking trustingly just as when we first met and discussed the project. He changed the subject from the bearings and opening his precious diary, read out some passages to me. One was very severe on me, but he turned the page and read another showing that he had found he was completely wrong and had changed his opinion. After that he often either read passages to me or gave me the diary to read what he had written. It gave me a new angle on the man and confirmed my suspicions that he was a newcomer to this part of the country.

Most of it was about the wrangling that had gone on among members of the party, then he wrote what he thought. Boiled down, it showed how suspicious he was of everyone. All through his writing it was plain that he trusted no one, not even the Sydney crowd, for it showed that he feared that once he revealed his reef, they and we would take charge, and not let him have a say in anything although all that was stipulated in the agreement. His writing showed that he expected and wanted evidently to be the manager and from this position he would dictate to the company. It was very plain that he did not intend to lose control, and this talk and reading his diary gave me some idea of what this little fellow was thinking.

In our agreement and in his agreement with the company it was stipulated that I would peg the leases in the name of the company, but tonight he pointed out to me that the agreement said that everything must be in his name and would remain so until the company had satisfied him that all his interest and cash options were covered. I did not argue with him over this, but knew that if we found the reef it would be pegged for the company, no matter how he read our agreement. All I said to this was, "I have no doubt that when you and I get to Alice Springs we will be able to cover everything you desire." He then asked whether he could depend on me to stand by him in everything, and I gave my word that I would stick to him and give him a fair deal.

"Well," he said, "if you do that, I will see that you don't lose by it in the end."

I have often wondered what would have happened. Our agreement with the company did not leave a loophole where any individual could take out leases in his name without the lease remaining the property of the company. I felt quite easy then on that point and could give him all the assurance he desired for no matter how or in whose names the leases were, they still belonged to the company. In the days to come it gave me food for thought and one day I actually found myself worrying about it for it was plain that Harry would put up a fight.

This was easily the most interesting evening I had on the trip. I had been beset by doubts and was inclined to worry about them, but that night the little fellow actually got me believing him again and once more I was full of hopes. He even came and gave me a hand to take off the camp ovens when time was up for the bread to be cooked. The others were in bed, but reading, for Phil always used the spotlight in our quarters at night. Harry told me not to say a word to them, but next day I had to explain, for they both knew something extraordinary had happened, when Harry and I were together for so long. However, I only told them that inside a week we would know if the reef was really there, as this was the country in which it was supposed to be.

Next day it was a very light-hearted party that moved on; in fact, at that period it was really nice. We all took the keenest interest in everything, even the cooking improved, or perhaps we enjoyed it more. We were passing through a lot of desert-oaks. We had been accompanied by them ever since we struck the sandhills, but these seemed to be a better type. They carried heavy foliage which made a dense shade. This desert-oak is a pretty tree to look at. When the foliage is thick it resembles a willow, and it is the largest tree that grows out on these elevated tablelands which are about 2,000 feet above sea level.

We saw many trees that were eighteen inches through with clean trunks to about fifteen feet up. There were no thick patches of them, but they were scattered everywhere. Sometimes we found about thirty of them in or near the breaks in the sandhills. In fact, we used these clumps of desert-oaks as a guide to the next break.

I had got up early that morning and gone over to where I had seen Rip Van Winkle's smoke. I wanted to see how many of them there were, also if they were camped on water. There was nothing, only the remains of a well-protected fire. I am doubtful if they slept there.

Just as we finished breakfast George saw his friend Rip coming up our track, almost running, and as soon as he got near enough, he started his crying, "Ningie, ningie", and making water-signs with his closed fists, putting them up to the side of his mouth. Then he made other hand-signs. George had saved all the scraps from the night before and what was left over from breakfast. We had tinned fruits and someone made a custard and we had too much. The same had happened with the porridge, so both of them were put in the scraps. There were two or three empty tins so we gave him these full of water and George asked me how he was going to give him all the scraps—the meat and bread was on top of the mush. We had nothing to give him it in so George tipped it into his old hat and, when the fool was going away, he tried to put the hat on and the porridge and custard ran down all over him. We got a great laugh out of it and thought the flies would buzz around now with all that sweet stuff over him. He was in a hurry so he carried the hat in one hand and the tins of water in the other. Phil tried to get a snap of him but the light was not good enough.

We had good, stony ground for the first three miles, then came on to a spinifex plain with big tussocks. These gave us no end of trouble; the differential gear-housing would pull them up but when there were several of them they stopped the truck and had to be dug out with a shovel. It was a brute of a job. In time we got wise to these big tussocks and dug out those that could be pulled up and laid them in between the others; this made a track so Phil could get about two hundred yards at a time. To give some idea of how slow it was, we only travelled six miles in four hours until the tussocks got thicker. This was fine, for once the wheels could get on top and stay there, the spinifex made a fine track. This lasted about six miles and we then ran on to really good country with scattered stunted timber. We were able to get into second and first gear on short stretches; we made fine time on this, and in two hours covered nearly twenty miles. We had been six and a half hours on our way; the first four hours we did six miles, the next two and a half hours we did twenty miles. We had lunch, for this to us was great going—twenty-six miles were already done and the country looked really good. There was a distinct change of landscape. It was the first slate country we had seen and looked as though it would carry mineral! It heartened us up tremendously, I can tell you.

17

A Possible Mineral Belt

IT WAS one of the most extraordinary coincidences that we encountered on this trip, that, as soon as Lasseter told me we were approaching where he thought his reef would be found, the character of the country changed completely.

When he made that statement to me, there wasn't the slightest indication of mineral-bearing country around or ahead and now we were running right into some where it certainly was possible. It fitted in so nicely with his story that George Sutherland, the only other miner in the party, and I began to think that in spite of all our doubts the little man knew more than we had given him credit for.

Both of us looked around at lunch-time and thought a reef could exist in such country. There were odd bits of quartz and ironstone lying about, and that afternoon we stopped to have a look at a likely-looking outcrop of stone.

Then we started climbing up a long grade. We must have gained an elevation of four or five hundred feet in four miles. From the crown of this rise we had a splendid view of the surrounding country, and could see an outcrop of razor-back hills. When we got opposite them we pulled up and inspected them. There were dozens of small quartz leaders running through them, and we decided to spend the rest of the day investigating. It was a good place to camp, plenty of dry wood and clean hard ground. Since we had got over the rise we had seen two smoke-signals go up, and when we were on top of the highest peak we could see smoke-signals going up from three places. These natives have wonderful eyesight for none of these smoke-signals was closer than five miles, yet as soon as we skylined ourselves, their keen eyes found us. George and I did a lot of stone-tapping and Lasseter also took a keen interest, since this fitted in grand with his talk last evening. We both told him it was likely looking country, and we dollied some stone, but the only mineral it contained was pyrites. We stayed up there until Phil called out that tea was ready, and we smiled as we went down, because this was the first time he had taken a hand at the cooking. The food was just as good as George's, and up till then George was considered the best cook of the party, although that does not mean he could hold a job down for long cooking at a roadside hotel. Those hotels, by the way, are supposed to be the tail-end of the world as far as cooking is concerned. When we complimented Phil on the job he made, he said, "I've never had a chance to show my hand. It's been nothing but punctures up to this."

That evening was the first time that mining was the main subject of conversation. George and I had both liked the look of this country. After lunch we walked about twenty-eight miles and saw that this possible mineral belt was over fifty miles wide. The trend of the belt looked to be north and south but the slate strata was east-west, so, discussing its possibilities, we decided to have another look in the morning. We had only examined one hill, and there were three others and about two miles of broken hills lower down. I went to bed that night thinking, almost dreaming: Was the tale Harry told me last night right after all? No sooner had he said that his bearing was about a hundred miles ahead than we came on to mineral-looking country and fifty miles of it. I was a bit puzzled by the fact that we were about fifty miles off Mt Marjorie but could not see it. The height given on the map was over 3,000 feet.

Next morning George and I went back to look at the other hills, while the others loaded the truck. It had been decided to leave a nine-gallon drum of water here so we buried it. I suppose it is still there. After we had buried it we dug another hole alongside and threw dirt on top of the hole with the drum in it, in the manner of a native grave. We did not do this to trick the natives but only in case one of the parties looking for the reef came this way. According to the Sydney crowd, and our Alice Springs mail, the West Australian party was heading for the Lake Macdonald country, which was the place we were making for.

We did not spend much time looking at these other hills, but came back to the truck, and told them to drive on for three miles while we had a look at the tumbled-down hills.

Harry said, "I would like to go with you."

I said, "Righto," so we inspected this lower country but saw nothing to hold our interest.

The natives showed much interest in our movements. As soon as the truck went off, up went the smoke-signals, then more again when they discovered us in the broken-down hills. These signals differed from the first lot as broken and short puffs were sent up; sometimes there was a cluster of four signals going up together. As far as the eye could see over that great field of sandhills on the plains there were smokes going up, evidently repeating the news of the day to other branches of the family.

The year before this these same natives were disturbed by two great birds that would fly over their country very high up in the sky. These birds made the same noise as this big slow thing that was creeping over the ground, leaving two deep tracks in the sand. This time the natives could see the people belonging to this strange monster; they looked like men but were all different colours. They had heavy-looking feet and two heads, one could even be removed—the first impression they get of a hat is that it is part of the man. No doubt these natives were puzzled and perturbed by this invasion. Their legends say that only the bad spirits cover their bodies. I have no doubt that excitement ran high. When we got back to the truck things were all quiet; not a sign of a smoke-signal anywhere; but five minutes after we started we could see new columns creeping into the sky.

The ground still held and we covered the next few miles in second gear although the grade was rising. Our aneroid showed 2,700 feet above sea level and right in front of us was Mt Marjorie but not so close as I expected it to be, for, in this clear, crisp air, one could see eighty miles quite distinctly and the mountain looked further away than that. Still, it did away with the need for taking our directions with the compass. Every time we took a reading it had meant that I had to leave the truck and get clear of the magnetic influence of our strong batteries.

Only half a mile ahead were piled-up sandhills and we drove up a little closer. Then the walking started in real earnest for I could not see an opening anywhere. Thinking I might be away a long time I took a waterbag with me. Harry asked me if I had my revolver, saying, "You never know these blacks. They are talking a lot with their smoke-signals and they seem to be everywhere." I told him I would not bother, as all the smoke-signals I had seen had seemed friendly. It was quite plain that they were only giving our position and direction. I wished I could make contact with them, for I felt pretty sure that this was their first glimpse of a white man.

I climbed the sand hillocks, but could not find a break anywhere. There was a high knob of rocks about two miles away so I made for that. It gave me a good close up view of this maze of tossed-up sandhills, and it seemed that we would have to follow the high ground and go north, for I could see far away where the sand was in long ridges. I went back to the truck. They had seen me coming so had put on the billies. I had been away for three hours.

After lunch we turned the truck round and went back to the top of the ridge. From there we could see the clump of timber where we had camped, about five miles away. We headed the truck north but soon came on to sandy going where the big sandhills were gradually forcing us to the north-east. After going about ten miles I took another walk, to see what the going was like over in a clear break, but I could not find a way into it. To use mats for over a mile to reach it, was out of the question, so we continued on our way; and by camp time I think we were only a little further away from Mt Marjorie than when we started in the morning.

I set some bread that night as I could see that I was in for a long walk before we moved the truck. The country was still providing interest for us, and although the slate had cut out, there were many small patches of broken quartz on the decomposed gravel.

We had a pleasant camp. A cold wind had sprung up so we rigged up a couple of tarpaulins and George and Phil sang a few songs. This was the first sing-song for about two weeks. I was invited to join in but I had done a lot of sand-walking that day and was enjoying a rest on the top of my bunk. I should say, enjoying a smoke, for I do not smoke when out walking; so I was refilling my half-dozen pipes and catching up some that I had missed and I was enjoying the singing for I felt happy.

I had had a short yarn with Harry and he was almost amiable. It was quite easy to get a laugh out of him. He was still a bit anxious about our direction and after a chat we turned in. I told him not to get up when I did, for I intended to go out early and explore the possibilities of getting through to the stony ground.

I was all ready to start out as soon as the light was good enough. I told George that the bread-dough looked good and ought to be ready in about an hour. Taking the waterbag I then set off and made for the open ground over the sandhills. It seemed about seven miles to me, but although I walked for four hours, I did not see a way through, so I had to return.

They saw me coming and put on a cup of tea. I sampled the new bread, then we all pushed on for about fifteen miles and had lunch on the way. Even on this ridge we were striking patches of sand that meant matting for a hundred yards or so.

We pulled up early in the afternoon. Our direction had been going away from the mountain and George and I again explored the possibilities of getting through a stony break. We walked right through it for about five miles and decided that we would attempt to get the truck through. It was very stony, with soft, drifting sand in between, and it would mean a lot of matting. We got back just before dark.

The natives had seen us and did a lot of talking on the air with their smokes, but so far we had not caught a glimpse of them, although that day we saw many fresh native tracks.

There was a fair amount of animal and bird life that told us there was plenty of water around somewhere, so we discussed the plans for tomorrow.

I felt sorry for poor old Mickey. He was really frightened of this country. Under our treatment his eyes were much better and he was watching the big smokes all day. Someone was always asking him what the smoke meant but did not get any change out of Mickey. His replies were always pretty much the same, "That feller talk along that feller smoke talk tell him we go alonga dis way"; but he said to me that night, "Big feller smoke talk about us everywhere me tink I sleepum closer up under truck, you letem?"

I said, "Yes, Mickey, you can sleep under the truck, you plenty wash and change your shirt."

He laughed and said, "Changem every day now." This meant he was using the three changes we had supplied him with.

Harry tried to talk to him, and asked him did he think they were bad blackfellows; but I had told Mickey early in the game I was his boss, and would do all the yabber for him and tell him what to do. So he put it over Harry by telling him to "talk alonga boss—him been no more". When Harry got back into the cab we all had a quiet chuckle about it.

It was very cold again that night and we had put up the canvas but we were in a poor wood camp and had very little fire, so all turned in early and did our talking from there. I think we were ten miles further away from the mountain that night than when we had started in the morning.

Next day we turned the truck straight into the sand, laying

down the matting from the word go; but before we had gone a quarter of a mile it was plain to us that this method was no good. There were odd clumps of spinifex about and we used these instead to get across the soft, shallow sand; but in spite of all our care the truck side-slipped off sloping rock and got down into deep sand where it bumped some concealed boulders very heavily. Twice this happened, then two driving wheels went down on one side; it was a bad jolt and it took us an hour to jack up that side and build grass and stones underneath it.

Later on, when we ran on to level ground, Phil made an inspection to see what damage had been done by those three hard bumps and found the main leaf of the big spring had broken. There was a clump of mulga not far away so we cut a six-inch log and put two blocks in between the springs. We had stocks and dies and a few bars of half-inch iron aboard, and Phil got to work making four U-bolts that fixed the trouble for the time being.

We were all day clearing the seven miles through the loose boulders, but all felt satisfied that we would have no more trouble there, for it was now a really good track.

We ran into a patch of stunted mulga which we had to cut down about a mile from where we got through the sand. Although it was only four o'clock we had all had quite enough, and made camp. Plenty of good wood was available, so this was a bread-baking night.

There were many smoke-signals in the morning, and I watched these very closely to see if I could pick up anything from them; for we had altered our course at right-angles. Sure enough the smokes altered, two short puffs then a long one. This, with other alterations, gave us some idea what the natives were talking about. In the main, they were only tick-tacking to others about our direction.

I paid close attention to the dawn smokes for this is the time they talk in figures. These smoke experts can send up a dotted line of smoke clouds that stand up thousands of feet in the sky. At this time of the morning the air is damp and cool and visibility is at its maximum. It would be quite possible to see these dotted lines a hundred miles away.

Although I did not understand the system I felt sure I would recognize any hostile or big call-up sign. When we got back on our western course the same signals were being used as before.

Mickey was useless with these smoke-signals. He had spent all his life under Archie Giles on a cattle run and had probably never made a smoke in his life.

18

Lasseter Tries to Take Bearings

NEXT MORNING was the most perfect clear, crisp day possible. Everyone felt it; there was a swing in all our movements. We packed the truck while breakfast was being prepared. By the look of the country ahead we had prospects of good driving all day. We had a bonny day, very little sand, but had to fix up a couple of creeks. The timber was very dense with long lanes and pockets. We managed to thread our way through without cutting a tree but we were still carrying our bumper bar.

It was a very easy day, no walking at all, and I think it was that evening that we ran on to a big, flat, granite rock. We stopped here, for it was a likely place for a good soak. Sure enough there was water in a big fault, or crack, a little lower down. It was a small affair, about seventy gallons, but alongside was a big swampy-looking flat that contained a big soak and this crack was being fed from it.

We called it a day and we were very satisfied. Our running had been about fifty miles. It was a lovely place to camp, and there was plenty of green grass and herbage, the shrubs and trees looked to be at their best. The trees all day had looked fresh and green.

Since we came into the "smoke" country we were paying more attention to our camping places, keeping in open areas where we could see two or three hundred yards all around, just in case we might be treated as trespassers. This was the first time we had camped on a waterhole. There were plenty of topknot pigeons around but as none of them were drinking and they all were flying to the north-east, we felt sure there were more waterholes.

We had plenty of time to have a look round before sundown. George and I went up to some big, round granite boulders and found another lot of water in a shallow pan four inches deep. It puzzled me to know why such shallow water had not evaporated, for there were signs at the edges that the water had been there for a long time. That night I got to thinking about it and decided to have another look in the morning—such a shallow water should not last for more than two days or so.

Phil gave his gun a try here; he was anxious to get more of these pigeons. There was also evidence of rock wallabies, but of course he did not get a shot; he had frightened everything away with the noise of the truck.

We put on two tins of water and Phil, George and I had a sluice down and changed our clothes, giving the used clothes a rinse in the water. Mickey also wet his other changes, just a lick of soap, but it freshened them up. I felt pleased at this because I had not told him to.

There was a slight wind and we all felt the cold very much and ate our tea with our top-coats on, all sitting round a good mulga-wood fire. By eight o'clock we were all bunked up, excepting Harry. He seemed to have a big write-up tonight in his diary. We had left the mineral-looking country behind as we had travelled miles in this granite country, but on our north side there were two big mountains, our map naming them Mt Flinders. I have seen three maps of this country and they all had a different name for these mountains.

Mt Flinders seemed to have a heavy growth of trees in the many small valleys that came down the rugged sides and the mountains were over 3,000 feet above sea level. Mt Marjorie looked only "just over there".

We had covered about a hundred and fifty miles from Ilbilba, but I am doubtful whether we were over a hundred miles as the crow flies.

During the next morning while Phil took a look under the truck, George and I inspected that shallow water. We came to the conclusion that it was being fed from a big patch of decomposed granite. When we came back later there was no alteration; and there was no sign of rain having fallen about here recently, although there were a number of these waterholes.

Phil reported that there were several broken leaves in the springs including the main leaf in the front spring. We repaired this; and then, because I did not want a hold-up until Harry had his bearings, we pushed ahead. We travelled fifteen miles that morning before we came to the end of the road in that direction. The country then for miles was all broken up and impossible to get a truck through. We spent the rest of the afternoon looking at every possible way, and Harry and I decided we would walk to Mt Marjorie next day. It looked only about four miles away.

It was a rotten camp; sandy, full of thorns, and windy. We had to rig up two of the tents to get out of the wind, but we were all of a tingle for tomorrow was to be the testing-out day and George and I were quite satisfied that there was no reef within fifty miles of the spot.

I was tired. I had done a lot of walking that afternoon, and the wind was blowing the sand everywhere. We all turned in.

The following morning Harry and I got away about 6.30. We reached the east end of the mountain by eight o'clock and, to our surprise, found the cliff was straight up; in fact the top layer looked to be overhanging and as hard as flint.

We edged round to the north side but I did not like our situation at all, for everywhere there were signs of fresh falls from this straight face of the cliff that towered 1,000 feet above us. We walked fully three miles before we came to a big crack, which we thought we might get up but when we climbed about five hundred feet, we came to the end. Harry proved a good climber, getting up about a hundred feet higher than I did, but he reported that a big overhanging shelf stopped him. I spent some very anxious minutes watching him climb down for the last ten feet were almost straight. The rocks were very loose and made footholds uncertain, so I asked Harry to hang on whilst I got up as far as I could and guided his feet into solid footings. He admitted that he had thought he would go over and was full of thanks for my assistance.

By the way, Harry very seldom thanked anybody and this lot of thanks surprised me. When we got down he said decidedly, "No more of that for me," and we walked for another mile before we found a way up.

This mountain sloped away to the west and was the shape of a wedge going up from nothing to about 2,000 feet in about four miles. All the mountains in this region are the same peculiar shape. It looked as though a section of the plain had been forced up, for the bottom layers of the mountains were composed of shale, the upper stratas were sandstone and reddish iron pigments up to the flintstone capping. I would think we had walked seven miles from the truck but soon forgot our hard climb—the view was something I shall always remember.

It was a brilliantly clear day and visibility was unlimited. We could see the Petermann Range quite plainly and it must have been two hundred miles away. Lake Macdonald looked a big affair even at that distance, about thirty miles, and the glare of the white saltpan hurt our eyes. This lake differs from the others as it had hundreds of smaller saltpans extending west to the skyline. Lots of others were a dull leadish colour and were either wet mud or water. A strange thing occurs on the wet mud shores of such lakes. I have found freshwater springs bubbling up. They are easy to pick out by the bird tracks around them, but all the same, great care must be taken of the ground around the shores. When the north-south line was being constructed along the shores of Lake Eyre men were lost through walking on concealed soft patches. All that was ever found was their hats. A stockman once told me that he saw his dog go completely out of sight in just a second in dry, hot weather; and there was little indication of these traps. Of course I did not go very near, so do not know whether these conditions exist at Lake Macdonald. You can bet your life I was not going near any salt lakes with the Thorny; its big weight would break through three or four feet of crust. I felt that thirty miles would be a safe margin.

Sixty miles to the north was a bigger mountain than this. There were other mountains all the same shape as Mt Marjorie, now so close. We had a clear view to the north of well over a hundred miles. The aneroid showed a rise of 2,000 feet since leaving the truck. As it was drawing near noon, we paid attention to the job we had come up here for. We were carrying three watches which read a few minutes to twelve o'clock, so Harry kept his eyes glued to the instruments. After a few minutes he said, "Now, have a look at that." I studied it for a second or two then said, "What am I supposed to see?"

He said, "You'll see half the sun." I could see that, but did not understand. Harry had done a lot of figuring after the shot with his sextant.

He said, "Well, I'm damned, if these figures are not the same as those we left sealed up in the vault of the bank in Sydney."

I said, "Do you mean that he took his bearings from up here?"

"Oh no," he replied. "Our watches were one hour fifteen minutes out when we got to Carnarvon, so seeing we were seven to eight hundred miles further west, I will try to figure it out. By my figures I have to allow one hour thirty-one minutes."

I said, "But you expected it to be one hour fifteen out, didn't you? Well what made you come up here? Was it just a desire to get as high as possible to take the shot?"

"Oh no," he said. "I had this all figured out in Sydney, that's why I was so particular in keeping our direction. I could have told you from the word go, that our bearings would be near Lake Macdonald but I had to come here to put myself right."

I said, "What do you want to do now?"

He said, "I want to get a hundred and fifty miles further south."

I replied, "When you do get there will it be where Harding took his bearing? That will take you in line with the Petermann Range. I have been close to that part and know that there is no mineral-bearing country out there. Had you said a hundred and fifty miles north, I know that is mineral-bearing country. Besides, it would fit in better with your Sydney story that you were six hundred miles from Carnarvon. By the time we get a hundred and fifty miles south of here, you will be nine hundred miles from Carnarvon! See if you haven't got twisted around somehow."

He peered over his figures and then did a bit more figuring and said, "It won't work, it's south right enough."

I said, "Well, old boy, we won't find your 'reef', for, close to where you want to go, there has been a big prospecting party out looking for a supposed 'big thing' like you talk about. I can reckon up twenty-two parties since 1908, looking for just the same kind of thing."

Then he asked me how I knew there was a mineral belt a hundred and fifty to two hundred miles north of us. I replied that I knew prospectors who had been out on it. It runs from the Western Australian goldfields right across through to Chillagoe. Of course there are big breaks in this belt, for quite a lot of it is covered by blacksoil plains. Tanami, the Granites, the Palmer and lots of other mining shows are part of this belt.

While we were eating our lunch, which was one tin of Queensland pineapple, I said, "Now look here, Harry, where did you get this tale from? Tell me everything, for the direction you now want to go is right into the big 'breakaway' country. So far, even camels have never crossed that country."

He replied, "I have known for some time that you have doubted my story but I want you to play me fair. I have no complaints up to date of your treatment. We have had our differences and some exchanges, but you are like myself, quick to forget, and I beg of you not to go into a lengthy statement of my story. I do admit some things were said to mislead you. Why I did that was because another member of the party had made me suspicious so I told little bits to mislead you, and I want you to trust me a little longer.

How many weeks have we before the hot weather comes in?"

I said, "Five or six at the outside."

"Well, give me that much time, for I am quite sure I will have enough indication by then to convince you that this is not a dream yarn. If the country south of here is as you say, then I would think there is something in your idea that I am working out these figures wrongly."

"Your distance all along has been five or six hundred miles from Carnarvon," I said. "There is a little cattle route over there," (pointing north) "about three hundred miles long, with a line of wells on it. I know two miners who sank those wells and they told me that some day when this country had a run of good seasons another Kalgoorlie would be found."

He said, "Well, it will be hard luck for me if I am turned around." So after a long talk about everything he asked me to give him another chance.

I said, "If you did the honest thing I would return and tell them that you failed to give me one single incident that was convincing. Your story does not knit up together."

Then he unfolded the map and said, "I did not intend to give you this information yet, but you see these hills marked the Three Sisters? Well, that's where I want to go for I am sure I can relocate myself once I get there."

I said, "It's time we got back." I scratched a message on the inside of the tin, also wrote something on paper, then put the tin in a crevice in the rocks. So if ever another pair of madmen go up there looking for a mythical reef they will know this is one of Lasseter's tracks.

19

Doubts of Lasseter's Story

WE DECIDED to go down by the south side and chance our luck there as we knew the north side meant four hours of hard going. We examined the country south before going down, and picked out a likely track. There was a steep gully and it was difficult to get a footing, and when we got down into the gully we found it ended abruptly in a big cave. I climbed up into it and was surprised to see its extent and felt inclined to examine it. On the dusty floor I could see plenty of fresh footprints so got out of it as quickly as I could. It is in such places that the natives keep their different kinds of wood carvings and their sacred writings. If one could only read their hieroglyphics no doubt we would find their full history written there. This is always very sacred ground and a good place to keep away from.

The undergrowth in the bottom of the gully made it impossible to walk there so we were compelled to get up on the rough side. This was very hard going, and, as it did not seem to improve, we decided to make a bee-line for the camp. When we got out on to the very stony ground we found two graves. They were remarkable; I had not seen their like before. There were about two tons of stone piled up around each mound. These stones were not just thrown, there was distinctly some order; the stones were in rows and kept fairly level. On top of the mound smaller stones were very closely packed; this was the shape of the grave as we know it. There were different stones around the edges and they were all water-worn. At one end there was a smooth large black stone, very flat and standing on its edge, no doubt some attempt at a headstone. Why I mention these graves is because this was the first attempt at stone-building or marking a grave which I had seen. We did not touch anything; I thought it probable that many pairs of eyes were watching us. The rough country offered many places of concealment.

We were a weary pair of men when we at last got on to the good walking. Phil had been watching our progress and met us with a waterbag. What I most appreciated was his thoughtfulness. He was afraid darkness would overtake us and as his truck was down behind a hill the spotlight would not have been any good to guide us.

I had a bit of a feed and then turned in, for I was greatly troubled. The show today had been that of a simple child. I could not pin Harry down to anything definite. I lay there and thought. I worked it out that a man who was supposed to have been here twice should be able to recognize the lay of the country from such elevated positions as we had been in. I felt certain by all his actions that this was Harry's first visit to this country. I had asked him to tell me what sort of country they crossed to get there and his replies were always unsatisfactory. They always ran much the same—"I was so excited at relocating the reef again I am afraid I did not pay much attention to the description of the country."

I had asked him in what direction was the water they had made back to, after finding the reef, and he supposed that it would be west or north-west. He told me they were fairly big rock-holes, but I told him there was no outcrop of rock in that breakaway country. He replied that not only was it a big outcrop of rock, but a big mountain of it. So, in my mind, I went over the day's talk.

I felt sure I would have been satisfied had he said 150 miles north, but the distances would not meet. He had said in the Sydney story 300 miles west of the Alice, 600 miles from Carnarvon. This left a gap of over 300 miles. I ended the worrying of it and decided to write down my views and findings and see what sort of evidence I could prepare. I had to make a report to send back to Sydney.

I would have gone to sleep with these thoughts, only Harry asked me a question. He had been writing in his diary ever since tea and asked me did I remember the figures of the reading we took that day. I answered him, "No, you put it all down in your book."

He replied that he must have lost that leaf, "I can't find it—

that's a darned nuisance, for it means I shall have to go out again tomorrow and take the reading again."

I said, "That's stiff luck," then all the doubts came flooding back. Here he was asking me for a set of figures that had worked out exactly the same as those he had placed in the city bank for safe keeping! He could remember that and yet asked me such a foolish question. I thought, "This will be sufficient to end the expedition", but realized I would have to wait and see what he did on the next day. He was evidently sparring for time.

After breakfast he got his things together and asked me if I was going. I said, "No, you can take Phil." Phil was pleased because he wanted to try out his new gun. Then I asked Mickey if he wanted a walkabout and let him go also.

To my surprise Harry said, "I need not go on top of the mountain. I think I will go on top of that hill over there, about three miles away."

After they had gone I said to George, "I wonder what his game is, climbing those hills to take his shot. Sea captains can take it from the deck of a ship; he need have gone only three hundred yards from the truck to the top of a knob and it would have given him an uninterrupted view of the horizon just as good as the hill-top."

George and I took a pick and examined a high face of conglomerated sandstone that was studded thickly with water-washed stones, very much stained with mineral salt. It was evidently from this sort of stony wash that the natives had got their water-worn stones for their graves. We soon returned to the truck. George threw his pack down exclaiming, "I'm afraid we won't blunt many picks on this job." I told him about the graves and we decided that the natives out here would be very intelligent as it was the first attempt at stone-building I had seen. We talked over the position, and I told George of my doubts. We went over things that had happened on the trip. George had asked Harry how he happened to get out to Cloncurry, for in those days it was a pretty rough place. "Oh," he had replied, "I just had the mining fever, so I thought I would hop in the train and have a look at the new copper field." George said, "You went out by train?"

"Yes, and a pretty rough trip it was too," said Lasseter, "but things were all excitement there over the new ruby find, so I thought I would go to the ruby fields."

George let it go at that, for Lasseter was talking of a time some years before the railway came to Cloncurry, and George had worked on the construction line. Later Harry told us that it was not Harding who found him perishing on the plains but some

Afghans who were out looking for sandalwood and it was they who handed him over to Harding.

At that time sandalwood was easily got along the coast, in fact, it had to be handy to cheap freight to be payable; so it was unlikely that anyone would be out 600 miles prospecting for that wood. Later on Harry told me that it really was a dingo hunter who had saved his life. When he made this statement I went off pop at him, and told him it was very unlikely that the West Australian Government was paying for dingo scalps in those days. He retorted that someone must have been paying for them, or these men would not have been out there.

I won't go on with the many slips he made. The discussion today was: Would we be justified in going any further? In our heart-to-heart talk on top of Mt Marjorie I had challenged him about these conflicting statements and he admitted that he had told fibs and said this had been done on purpose, but that the Harding story was the correct one. "I only told you those yarns because you would not believe in me and I thought you might turn it in. Then I could go out with camels, for I had been told in Alice Springs by one of the party that you had put Colson on so he could get in first and register the best of the claims. From then on until lately I did not trust you much."

All that day George and I thrashed the matter out and came to the conclusion that Harry could put up an excuse and probably get away with it by saying that he had thought that I would double-cross the company by working in with Colson. I had by now received a letter from Sydney telling me to sign Colson on as the company was afraid he would be able to stake claims for himself. I replied by saying that their fears were groundless for Colson would not be with the main party; his job was to bring up supplies and he would always be behind.

When I thrashed this out with Lasseter he again had excuses and said, "We were afraid that you and Colson would work together as Colson would not be bound by the company agreement and could have a free hand." We both knew whom he meant by "we". Before Harry and Phil returned George and I decided to see this "a hundred and fifty miles south" stunt through, for Harry was very difficult to pin down to a statement that would go on paper.

When I had decided to go on with the game I was in a much better frame of mind. Up till then I had borne all these doubts alone but now felt much relieved after I had unburdened them to George.

When Harry came back he told us that he was more convinced

than ever that the reef was to the south. I asked him what difference was between today's and yesterday's shot and he said practically none.

After lunch we turned the truck and went back to the big flat rock where the waterhole was. The fifteen miles run back took only an hour, and all got busy unloading the truck, then jacked-up the broken spring side. Also the main spring on the front wheel had to be attended to. It was a big job to go over all the springs and it took all the next day to make them as good as ever.

This was where Phil showed his skill as an engineer, for he removed the broken main spring of the back and cut and fastened it to fit the front. He did the same with the leaves, as the big spring took all our spares.

It was four o'clock by the time we had everything fixed, and, as we knew there was extra good going for the next twenty miles, we pushed off. By 5.30 we had covered thirty-one miles, plenty of timber and turpentine bush. We decided to send up a big smoke in case Fred was close handy and at night played the spotlight for we were expecting Colson any day now.

20

Back to Ilbilba

AT DAYBREAK we decided to go back to where we had turned north and see if there was a way of getting over towards Lake Macdonald country. That run back was easy. We arrived there in two and a half hours; having the road broken made a lot of difference. It had become very warm so as soon as we came to our east-west track I took a bit of a snack and the waterbag with me and set off to see what the flat country beyond the sandhills was like.

I followed a channel between two sand-ridges; it was heavy going but not impossible, so I continued for eight or nine miles to the open country. I only walked a few feet on it and decided that it was impossible for the truck. It was soft soda and salt ground, anything could be under it. Besides, there appeared to be rough ground ahead and it appeared very much like a continuation of the breakaway country I knew of three hundred miles further south. I returned just before dark and told them it was almost impossible; the clear country that we could see was soft and might be part of Lake Macdonald. So that night we decided to go back to Ilbilba and try to force a passage south.

We had a splendid run next day back to Ilbilba but had a new annoyance—a following wind which was quite hot. Our radiator was boiling all the time. It gave us a taste of what it would be like in a month's time. The radiator used thirty-three gallons of water that day so I had a talk with Phil about it and he said he would have a try at fixing a condenser. We had all the necessary piping, for we carried a condensing plant.

After we had a look around the bough sheds, Phil, George and I took the tub up to the water and had a good bath. The sun had set before we got back but there was quite a warm wind blowing. Harry had cooked a real good feed and we enjoyed it. Harry then asked if he could take out the receiving set as he wanted to listen in to any news. He was anxious, he said, to pick up something about the parties that were out looking for the reef. We all had a bit of a listen in but I thought it was too much bother sitting there with earphones on. Mickey had a go and enjoyed it immensely. He'd listen to music as long as you'd let him and he especially loved the concertina—Archie Giles played one!

When I was making for the bough shed Mickey asked me if he could sleep under the truck. I said there was nothing to be afraid of; there were no tracks here anywhere, but he replied, "I been seeum plenty of smoke today, that way," (pointing south-west) "last time him close up along here." So I told Harry that Mickey would be sleeping under the back of the truck. This had been his position ever since he had left here and he still wanted to have it at Ilbilba. Phil wanted to know what Mickey had to talk about. I told him he had the wind up and expected a lot of natives to arrive. Phil was quite concerned and wondered if it was wise for us to be so far away as we were sleeping a hundred feet or more from the truck.

George woke up before daylight and, out of devilment, made a lot of tracks with his naked feet. Soon after dawn the flies came and that meant the end of my sleep, so we all lit up and were talking and smoking when George "found" the tracks. He sat up and said, "Old Mickey was right, look, they've been right in here."

Poor old Phil was completely taken in for a while; he got up to examine them closely and to see if they had been over to the truck but he tumbled to the joke when he saw the tracks only led to and from George's bunk. We had a great laugh about it.

There was nothing to do on the truck now that Phil had fixed up a condenser, so we three, and Mickey, took some more tins up to the waterhole and all had a bath and washed our clothes. Phil shot five pigeons and after lunch George and I took a walk around the hills but there was no sign of any mineral. We found many tracks of a strange bird; it had only three toes like an emu and each only one inch long. We tried to find the birds but saw nothing; which was just as well, because later I found out they were sacred birds.

We got a good dinner of stewed pigeon that night, Phil had bagged some more in the afternoon so they were put on to cook for breakfast. We had not had a feed of fresh meat since we came back from the wrecked plane, when Fred had brought out some steak.

When morning came Harry told me he would like to go out over the range and inspect a big valley. I said, "All right, you can take Mickey along with you." So off they went, but Mickey would not go over the range. He came back to George and Phil who were trying to get more pigeons.

I asked Mickey why he had left Lasseter. He said, "Too many wild blackfellow plenty make sign talk about us." I saw he was badly scared. White people often get their first glimpse of real terror when they see a native's fear of his enemy brethren. I felt a bit concerned about Harry then as I was doubtful of his bushcraft and his ability to get back by himself and questioned Mickey closely. Mickey was quite reassuring and said, "Him come back all right. Him make plenty white track alonga stones with white ochre."

Later that evening Harry returned with several pieces of stone, mostly rock of no mineral value. I asked him to pick up any likely specimens he might run across—chiefly to test his knowledge of prospecting on mineral-bearing rock.

As I was curious about these white tracks Mickey told me of I took him out with me the following morning and he showed me where Harry had been making marks with chalk on the stones to be certain he could find the same way back.

Lasseter's version of the walkabout was that he had walked over the range about ten miles into a big valley. I felt very doubtful of this—he did not appear to be tired enough after his thirty-mile walk—so, with Mickey's aid I tracked him to a clump of trees about five miles from the truck where he had evidently put in the day and had a good rest. That, to me, was pretty convincing proof that Lasseter had not had much bush experience.

I had agreed to stay around the truck that day; we never left it alone too long. During the day I did the meanest thing I have ever done in my life—I opened Harry's tin trunk and went through all his papers, also the box with some gold specimens. I think there were seven stones in it all showing gold freely. I got out Phil's breast-drill and bored a hole in each and filled the hole up with white metal, concealing the hole with glue. I did this so that if ever the stones turned up again I could identify them. My main reason for opening his trunk was to see that Captain Houston's book on navigation was still there. It had not been opened since the captain had used it at the Dashwood. I found plenty of food for thought in that trunk. Seeing that there was such important evidence in it, it's a wonder he went away and left it unlocked. This was the first time I had ever seen it without the lock and I wondered if he left it like that often. I tried to read, but a golden hornet was buzzing around.

By the time the others got back, Old Man Worry had taken a firm hold of me, for Freddy was now four or five days overdue. While we were cooking the pigeons I told George and Phil that the waiting was as bad as waiting for the mad agreement to expire. Both assured me there was nothing to worry about because we had agreed to allow three or four days' margin.

When Harry came back tea was ready but I noticed that the first thing he did was to open his trunk. Afterwards I went over and saw that the lock was on it again so that leaving it open was evidently an accident. By the way, we had to wait until after dark before we could eat pigeon and drink broth, for the flies were here in tens of thousands. It was nothing to have five flies drop into your pint of tea in daylight.

I asked Harry to come into the bough shed that evening and discuss what was the best thing to do but I was ruled out of order again. I wanted to go on again next day; but George and in fact all of them explained that we had plenty of tucker and water so why all the worry? I told them I would not go on again until I knew for sure that the auxiliary truck was coming for if anything happened to our turnout, our existence would depend on being relieved by it. The fire danger was very real, more especially now that warm weather had come in. We always carried the fire extinguisher on our knees when travelling. We packed drums of water and provisions right at the back in case the fire got out of hand. That is why I was so anxious to find out about Freddy. Should a fire get out of hand we might be able to pull the water drums off but we carried a lot of petrol besides the explosives, also the gas cylinders of the oxy-welding plant. A good place to keep away from once a fire got out of hand. Without Freddy I could not go a yard, as I considered I would be meddling too closely with the Hereafter.

That afternoon Rip Van Winkle came in with three women and six children. Of course Rip came right over to us and started his whining "Ningie ningie". The women had sat down about a hundred yards away. We went over to have a close-up view and the lady we had seen before gave us a nice smile uttering two words that sounded like "pin dar". There was a much older woman, and one much younger, about fourteen years of age, but fully developed and one could say fully grown. She was a picture to look at. I did not think that she had had a child.

The children interested me more. The youngest baby was a hideous thing to look at; it had a big head that rolled about as though its head was rubber, pulled-down eyelids and practically no nose and a slightly protruding tongue, a small round body that seemed to be all stomach. The arms and legs were long and jointless and their slow movements reminded one of long spider legs as the child lay there on the sand. The sight of it nearly made me sick. The other two children seemed very bright and on the point of laughing all the time, but even one of these had a paralysed arm, so out of the six children there were only two that were not crippled.

We could not make any sense out of them at all, but it was quite plain that they had contacted whites before. I tried to talk by signs to the one we had first met for she looked very intelligent, and I think if I could have got her away from Rip I could have made her understand, and, maybe, have understood her; but this old thing with his old felt hat would chip in with his maddening cry. Then the bright kiddies started talking and pointing and another male native came in sight. He was very tall, about six feet two inches, with a bald head. They had no weapons. The women had three nice dugouts about two feet long and six to eight inches wide, about seven inches deep. They were a work of art for they had been pared down to only half an inch in thickness. The bowl would hold about two gallons of water and the wood looked to be beanwood.

The women always had yam sticks, a sharp broad-pointed stick used for digging yams or native potatoes. We did not give them any food, just walked away. Rip made signs to me later so I went over to him. He took me over to the young woman and quite plainly offered her to me; then I suspected this old sinner of being able to talk. Later on I took some tucker over to the women and children because I did not want the men to eat the lot. I got a shock when I saw the young girl suckling the monstrosity for I had thought by her firm breasts that she was not yet a mother.

I gave all the kiddies a good handout, then the women. The two men looked disgusted, since their law is that the men take what they want first, but they got a good handout too. Mickey had a good look at them and I told him to speak to the women. He tried two or three times but they seemed afraid of him. Mick looked as black as charcoal alongside these people.

Old Rip and his bald-headed friend seemed to be always watching the west; their women also seemed anxious. About an hour before sundown they moved off towards the waterhole. I asked Mick if he could make out what they were looking for and he said, "Him been watchem plenty smallfeller smoke—I tink plenty more blackfeller come alonga."

So next morning I asked the others not to go away until we saw what this smoke yabber meant, for it was quite plain that Rip and his tribe would not have left good easy tucker if something were not wrong. There was no cover for natives to use, only from the gully there were fairly dense clumps of bushes just off our track to the waterhole, so I put in my time watching in that direction.

I suppose it was about eight o'clock when I saw a head come above the bushes. I felt that this would not be Rip but would be the men who had been sending the smoke-signals the day before. All of us could see him watching so I said that I would go half-way and give him a sit-down signal. I walked out and gave my signal and to my surprise the native stood up and seemed to be talking to others. He came forward a bit but still in the bushes. Then I gave him some sign-talk by holding up both hands, then lifted each foot. He came forward. It was fine to see that stately fellow advancing at a fairly quick walk. Of course I said, "Good day" as soon as we came close together, for when he started walking towards me I walked towards him; but what he said I don't know. He talked a good deal, pointing west, then south-west. Then I gave him a water-sign. He readily understood, pointing towards the waterhole. I was pleased at this for the sit-down and the water-signs belonged to a tribe four hundred miles further south. Then we got to sand-maps. I cleared a place, made our truck-track pointing out west. Then I made another track and pointed south-west, then gave my water-sign.

He studied my plan for a little while then made signs on the sand-map, but I could not understand anything. He saw this so I stood up and gave a few passes with both hands, then four other men stood up. He gave a one-handed sign very much like my sit-down sign, then these men seemed to be doing something with their feet. They all came towards us and as soon as they got clear of the bushes I could see two children, one a very small chap. They came right up without talking, the little boys sitting down about a hundred yards away.

Then the first one started explaining my sand-map. Each one at once cleared a place, sitting on his haunches. They drew a plan; then the first touched my plan, then pointed to each of their plans. Quite a simple explanation according to the look on their faces. I thought Mickey might understand them so I pointed towards our camp, beckoning to each one. I turned and they followed but would not go nearer than about fifty yards, so I called Mick. I could see fear written all over his face. I explained to him to try and understand so we got to work on the sand-maps again but all that Mickey could make out was that these came from the south or south-west, for on the ground they were explaining that their tribe was over there; but they would not respond to my water-sign in their country, they always pointed to our waterhole.

These men were fine to look at. Two of them would be over forty years. The leader, who came first, had a few grey hairs, but a nice open face. The easiest way to describe it would be to say it resembled the familiar face of Daddy Xmas. The others had beards, but the young men were only growing a few hairs on their lips and chins. Their bodies were clean, their hair was free and well kept; no sign of dirt at all. The leader had hair about six inches long, the others much shorter. They were lighter in colour than Rip Van Winkle, but I think that was only because these chaps were clean. They were only a shade darker than my sunburnt arms. Each one had some sort of charcoal marking on his body, the charcoal showing up quite distinctly. I noticed that the two older men kept running their fingers through their beards. Their beards had quite a distinct wave. Mick, in colour contrast to them, looked really black. These men were all circumcised and three of them had a heavy fold of skin or muscle coming down in a V-shape below the navel.

By this time the rest of the party was standing in front of them. They showed more interest in Phil than any of us; he had on shorts and a sports singlet and was burnt almost as dark as these fellows. They walked around him and had a good look at him. Phil had a fair amount of black hair on his legs and this seemed to cause much amusement; they had clean legs and only very few hairs on the chest. Their hair was a dark brown with a good sheen on it. The two boys were fine little chaps; the elder one, about nine years of age, showed a marked intelligence. The boy of four years was very shy for a start and clung to the leader. I found out later that he was the boy's father. They were both of very fine physique and well proportioned. The men were sturdy in development, the shortest one was about five feet seven inches and the tallest about five feet nine inches. If science had the fashioning of man it would use these men as a pattern.

We had a tin of lollies, so gave the kiddies a handful each. I had to show them what they were for. I took one and put it in my mouth and the little fellow slowly put one in his mouth, at the same time watching his father. It was great to see the surprised look on his face when he tasted it.

I then did the same with the big boy; so far he had not smiled. Taking one of the lollies I had given him I put it in my mouth and opened my mouth so he could see it in there. He seemed to be more concerned with the second eldest man. After a bit of coaxing he put one in his mouth. I only wished that I had a movie camera to capture the change on his face. He had rather a sullen expression, but when he smiled it was the most wonderful change of expression I have ever seen. Had a film company snapped it they would have made it their slogan picture, for I don't think an audience would ever get tired of that change of expression.

The men seemed to enjoy the children's pleasure but none of them attempted to take a lolly.

It was only a couple of hours before both boys were mingling with us. The big truck seemed to fascinate them. They were interested in the wheels and both put their fingers in the tread of the tyres and yabbered away to each other, the big chap pointing to the ground. The little chap would poke his finger in the imprint and both looked along the track of the wheels, pointing the way the truck had come. They had evidently seen our tracks and seemed quite delighted to find out what made the track. Phil lifted the bonnet to show them the engine but the part they were interested in was the bonnet's movement; it seemed to please them.

I doubted if these men had ever come into contact with whites before; they did not understand the use of matches. Of course they were very interested in them the first time I lit my pipe. They all crowded around but they were not amused, their expression was more one of disapproval. But they were interested in the used match; one of them picked it up, and each examined it closely. I struck three or four matches just to show them and offered them the box, but none would touch it.

Phil got out some petrol and poured it over the sand, then acted the fool like a magician does, passing his hands over it; then he lit it. Of course there were flames two feet high. I watched the men very closely. Their expression changed to one of surliness and they kept muttering as though they were annoyed; in fact I thought they were going to clear out, so I kicked some sand over the fire. Of course it went out. The leader came over and examined the ground closely, putting his fingers on the place that had been alight, but we could see his nostrils working; he was sniffing the fumes as though trying to recognize the odour.

Phil or George then gave Bubbles (that was the name we had given the boy) a looking-glass. I think I got more amusement out of this than anyone. He looked into it and laughed then turned it round; it was great to see the surprised look on the kid's face. He turned it round, saw his own reflection and laughed again. His laugh was grand to listen to, such a hearty sound, full of bubbling joy; but when he started looking close into the glass, I nearly died. He was holding it edge-on trying to keep one eye on the reflection and one eye on the back. He would turn it around and around, not believing that it could disappear so quickly. Then he started using his brains. He held out the glass in one hand and was laughing just as white children do when they are about to expose something such as finding the other child who has been hiding away. Bubbles's laugh was just like that and he put his hand around the back of the glass fully expecting to catch hold of the other kid, but when he could not feel anything the look of astonishment was grand. He turned the glass over, then looked on the ground and behind him. I could see he was frightened, so I took the glass. The men looked very displeased and came and took the boy by the hand. I then saw that we must use a bit of sense if these people were going to be of any use to us, for it was quite plain that we had gone too far.

Lunchtime came and we gave them bread and butter and meat sandwiches. They hadn't the faintest idea what to do with it—clear proof that this was their first handout. I had to take a bite out of the leader's sandwich and show him what it was; soon they were all eating. After a while I showed them flour and rice which they did not know. I got some empty fruit-tins and put them on the fire. As soon as they knew they were being fed they lit a fire, even though the sun was still hot, but they made the fire in the shade. George reckoned they made the fire to keep the flies away while eating.

They showed more interest in boiling water than anything else we had shown them. One of the young men put his fingers into the boiling water. Of course it burnt him and you should have seen the ugly scowls on their faces. I saw at once that this was a mistake and that these men would resent any tricks being played on them so I put my finger into it quickly, then into some cold water. I did this two or three times. Then the young fellow with the burnt finger put it into the cold water. That was all right and he yabbered quite happily, quite a lengthy statement, pointing to the hot and then to the cold water.

When the rice was cooked the trouble was how were they going to eat it? We could not give them our spoons for they might not want to give them back, so I got the snips and cut up a tin and tried to fashion out a spoon. Phil soon got fed-up at my attempt at tin-snipping and took the tin and soon had seven spoons made. When the rice was cooked we put the tins on one side to cool then showed them how to eat it with a spoon. Both of the kiddies soon grasped the idea.

Incidentally, both of the boys could repeat any word you said to them. It was grand to hear them repeat the names. We called the big one Charlie and he would drawl the word out in a slow musical tune. The little boy cut his name short to "Publs".

21

We Meet the Dogger

ABOUT TWO O'CLOCK we saw the natives looking up towards the waterhole and we saw three men and five camels coming. "Gee!" I said, "This is visiting day at Ilbilba right enough."

Harry immediately said, "This is one of the parties out looking for the reef." When they got a bit closer we could see it was a white man with two natives; all were dressed.

After greetings the white man said, "I was surprised to see fresh tracks up at the waterhole, so came along to see what all these truck-marks meant. I'm pretty short of tucker and thought you might be able to let me have some." I asked him what he was doing out here and he told me that he was hunting dingoes; he had been out four months.

I questioned him about other parties but he was not aware of any goldrush. While we were boiling the billy each one asked him questions. He and his natives were on their way back and had been forced in here for water. I soon found that one of the boys could talk perfect English. I gave him tucker and tea and sat apart with them questioning him, but was quite satisfied they did not know of our existence. The white man called himself Paul Johns; he was a German of about twenty-two years of age. He told us that he had only been in Australia three years. I remarked that he spoke very good English and he replied that he did not know many words when he had come but the Mission had taught him. He said that he was considered well educated in Germany and that helped him to pick up English.

The men were hungry; they ate a loaf of bread and a tin of meat each, besides some cake.

Johns said, "I was very surprised when I saw all the water-drums at the soak." (We had taken the precaution to fill several in case this water gave out when the hot weather came in.) "I lost no time in getting down here, since we are practically out of tucker. We've been living on kangaroo." He asked us if he could have enough to get him back to the Mission. We told him he could have what provisions he wanted, also tobacco, and made a bargain with him to stay about here a while. In two or three weeks' time I would know whether we could get through to our position and, if we could not, I thought I might hire him and his camels to carry on.

He readily agreed to do this and hoped that we would give him a job, as he had not had one in Australia.

I told him of the many dog-tracks we had seen out at Mt Marjorie and Mt Flinders and we told him of the two waters we had found and said that, judging by the flight of the birds, there were bigger holes to the north and north-east. So a bargain was made for us to supply him with tucker while he waited and together we made out a chart giving the dates when we would send up a signal from the highest western point on the Ehrenberg range, about five miles from Ilbilba, fixing the time at 8 p.m. We decided on twenty-minute flares and rockets. He agreed to get within sight of those points on those dates. I thought it a great stroke of luck and saw a way out of our difficulties, because with these five camels on call it meant that if we got into country we could not get the truck through we could use the truck as a sub-base and get almost anywhere. I said, "Your boy can talk remarkably good English."

He said, "Yes, he was educated at the Mission, he can read and write fairly well, too. He can also read and write and talk German. The Hermannsburg is a German mission station and he is one of the Aboriginals who can read and write in two languages, besides knowing several tribal languages."

And yet some people think that the Australian native is not of high intelligence! In my opinion it is the white man who needs to rethink the matter. How many of these so-called cultured people can talk half a dozen languages, to say nothing of sign, smoke and tapping languages? Yet here is one boy who can live as his people do, with a spear and stone axe, or can sit down to any table and eat correctly.

This native should be employed by the Crown as an interpreter in Aboriginal trials, then our natives would get a fair trial. I admit that our courts of justice would want some reconstruction, since as constituted all they think of is punishing the natives. Some even want to flog them. Compare the intelligence of this native with theirs.

I said, "Does he understand the language of this tribe here?" He said, "A little; he knows the language of their women, for he has taken one of the girls for a wife."

This seemed even better. I had a talk with Ralph—that was his name—and asked him would he come over to the five natives and see if he could find out something about the country we were making for. After they did a lot of talking with many sand-maps he told me that they only understood a few words, since this was a different section from that where he got his wife. Afterwards he told me that they would not tell him if there was any water in our direction. He told us that they did not want us to go out that way. He said, "These are light-coloured people, they belong to the sand-hills country." I asked him how many tribes there were of them, but he did not know for sure. The Mission classed them as three tribes but he said there are at least seven tribes but did not know if they inter-married.

I asked him whether the people of the spring country belong to the circle of changing women. He was not sure but thought not as they were of a darker colour. He said that the people of the sand-hills were very wise people and were very friendly. They did not fight other tribes and they worshipped some sort of god. They differed from other tribes because the father took the boys away from their mothers when they were very young and though always together in a tribe or band these boys never returned to their mothers. They were taught the men's language and the craft of hunting or throwing spears and the making of them.

I said that they were a very fine-looking race of people and asked what their women were like. He said that the sandhill women were the best of any around the Mission. He had often heard the missionaries say that they were very beautiful. I said that it was a wonder the Mission did not go out there. He said that they tried but the sandhill men would not have anything to do with them and the spring people showed fight. I asked him whether he had ever seen any of the sandhill women but he said only the married women and they did not belong there but came from other tribes. These tribes inter-married and practised circumcision; that's how they tell the tribal boundaries. No man is allowed to look at the virgin women until they are ready to marry.

I learned from Johns that he and his boy helped to clear

Mackay's aerodrome. They were here at Ilbilba for three weeks with Bob Buck's party.

I asked him if he knew Rip Van Winkle's tribe and he said he knew the tribe I was talking about but they were bad people. The men were considered to be mad and both the young women were the daughters of the older one and both of them had children to their father, meaning Rip Van Winkle.

I said, "Where does the tall bald-headed chap come in?"

He said that he came only last year and that he did not know where he came from. I asked him whether he had seen the children and he said yes; the four crippled children belonged to the daughters of Rip (he called him by another name). The two bigger children belonged to the mother; he did not think that either of the daughters was over twenty years of age. I asked him if he thought Rip could understand English and he said he thought not. They were there when he was building the bough sheds and clearing the drome. Buck always talked in native language and they had helped him clear the land for the aerodrome.

Ralph was a very religious boy and thought it a great sin for Rip to be allowed to go on like that. I said that it was a wonder the sandhill men did not kill him, but he said, "No, no the sandhill men do not kill, they threw him out of the tribe." I asked what tribe Rip belonged to and he pointed to the north-east and said that he belonged to that tribe there. They were only a small tribe, so that was the history of Rip. He was an outcast, evidently a sexual maniac. I remarked that he was circumcised and Ralph said, "Yes, that is to determine the tribal boundaries, these people of the sandhills are all the same religion."

In no other part of Australia is there anything like this, for as far as I know no other tribe worships a god. In fact, none of them seems to have any fixed ideas about religion; yet these people of the sandhills say prayers. I asked Ralph if he had ever heard of a Jewish rabbi out here and he said that he had heard of him, as a boy, but never knew what he was doing here.

I had a lengthy letter published in the *Sydney Morning Herald* in 1932 on this country and in it I advocated making all the sandhills country a native reserve. It brought a reply from Mr Bridges who told me in his letter of this Jew. I think he described him as a Rabbi scientist who spent two years somewhere on the fringe of the sand belt studying these people. He discovered that they used nineteen ancient Jewish words in their ordinary everyday language. I think that the Commonwealth Government should trace this Jewish scientist and get his report of what he found, because to my way of thinking this document could be important as the first step to understanding and finding out who these sandhill men are. It should not be hard; maybe Mr Bridges could tell the Government where to look for this very precious document.

The sandhills should be restored to these people, land that will never be any good to the white man. Give these tribes about thirty good surface waterholes and they will soon multiply. Make it a reserve in every sense of the word and police it so that the wandering outcast, the Combo, the sexual maniac who is responsible for seventy-five per cent of half-castes, cannot get in.

At the bough sheds, night had fallen and there was peace in Ilbilba that gave a wonderful feeling of complete relaxation. The mission boys were singing hymns with beautiful expression and time.

As I sat there listening to their song I couldn't help comparing the separation of time, for there on one side was the product of man's inventive genius, the big truck and on the other hand a large aerodrome that had accommodated the dreams and brains of the world—the flying machine. There sat five white men, two from far-off countries; they were watching and listening to the melody of song of early man who was little different tonight from when time began—all these stages of man, yet one section had not altered. It had remained as God had fashioned it. Compare the five sandhills men with the five white men. Their life of peace, happiness and freedom. Then the five later products of civilization. Who are the happiest and freest? Then the three natives with white man's covering. They had lost everything. Lost the age-old man's happiness and freedom and gained nothing of the white man's pleasures. Evolution had only brought misery to these white men's slaves.

Yet separation of time had made no difference tonight; we were all equal. Both ends of time were listening and enjoying the mission boys' beautiful song. Then it changed. A new voice, Charlie, the child of the sandhills, had caught the air as it went along. It became stronger as it developed into a sweet contralto. I shall never forget those few bars with that clear, sweet voice. Then, the marvel of all, the five white men sat in a circle with early man and joined in the singing led by the mission boy, Ralph. The young German had a good voice, it was strong and the others following in gave the melody increased strength, echoing over the rolling sand.

22

A Hot Report to the Sydney Crowd

NEXT MORNING at peep-of-day activity began. We had decided the night before that we could go in and find out what had happened to our auxiliary truck. With the natives all fed, we were ready to start and we said goodbye to Johns and wished him good luck. They went west and we continued east. So over the sandhills once more to pick up Colson, who was overdue. For the second time the old truck was on rescue work.

It was not palatable to any of us, but in the circumstances there was nothing else to do. This was the fourth time the big truck had gone over the road. The truck ran easily. We had no stops and averaged about fifteen miles an hour. We just sat there in the heat of the cab. The radiator was boiling a treat, but Phil's new condenser was wonderful. We were all very silent and felt how useless the trip was. We pulled up late in the afternoon at Taylor's Creek to boil the billy, for we had not stopped for lunch. We had millions of flies with us and no one felt like trying to eat. Harry went over and had a look at the marks where the plane had crashed while Phil and I ran off the boiling water in the radiator.

George said, "Tea is poured out. Come quickly or it will be full of flies." So we all sipped the good, hot tea. It is the best thirst-quencher I know of. Phil said, "Come on, all aboard," and away we went.

We had to get out the matting to get over the three sandhills and the creek but it took only five minutes, then on, for if we did not meet Freddy we would be in Alice Springs the next night. What a contrast to breaking and making the road. It took three weeks to get out; now we were going back in two long days. Phil asked how far we would go that night and I told him it was up to him for we were only passengers with nothing else to do but hunt flies out of our eyes.

"All right," he said, "we will go to that bad creek with the angle crossing."

We arrived there after sundown. All daylight had gone before we matted the truck across, so Phil called it a day. On this side it was hard, smooth ground. We all set about getting a feed ready. George made our favourite, a dry hash stew. This is made by mashing up the potatoes with butter, then adding four or five onions cut up very fine, then mashed up together with two tins of bully beef. With no flies to trouble us we were all enjoying the feed when we heard the drone of a motor. We all said together, "Here he comes." I looked in the pot, yes, there was plenty there; we knew he would be as hungry as we were.

In about twenty minutes he pulled in. Of course, we were all very pleased to see his old smiling face and to know that he was all right. The first question was, what had happened.

He said, "When I got back with the plane I started on my job of fixing up things for my wife as I did not know when I would be back. In about three days I was all set to take on the last of the petrol when I received word from the Sydney crowd that they did not require my services any longer."

I said, "Isn't Coote in the Alice?"

"Yes, but I think he knew before I did. I got busy on the wires and told the Sydney crowd that the party had gone out west and depended on me for supplies." (I had told Colson before he left me that if the indications were good enough I would go beyond the safety margin and therefore I would not be able to get back without his assistance.)

He said, "I told Coote this also but nothing that I could say was of any use. Finally the Sydney crowd closed the argument by telling me that I was sacked, and would not be required any more. The Sydney crowd had told me they were running the expedition and would manage as they saw fit. I then went to Mr Allchurch, the magistrate, and told him of the position that had been created. After he had consulted with the Government Resident it was decided to do nothing until they found out if another truck was going out. They asked how long the party would be before they were in any danger and he said, if things went all right, about three weeks, but if the truck caught on fire it would depend on how much they rescued as to how long they could exist. I told them every precaution was being taken against fire but one never knew when a drum of petrol would spring a leak. In that event very little of anything could be saved. So they decided to wait a few days and they would find out whether another truck was being sent out."

Fred had been trying to get a loading for a station, a short trip which would occupy only three days or so. He secured the job, and drove out with the load. In the meantime, Mr Allchurch had found that no other truck was going out but that the Sydney crowd was sending another plane.

"When I got back I saw Coote," Fred continued, "and told him I was going out on my own initiative and would make a fast run and tell the party what had happened. I then informed the magistrate and he thought it was the only thing I could do.

The magistrate said, "I am glad, for I certainly would have taken the responsibility of sending you out to inform them of the altered conditions."

Colson added, "I intended to come out, no matter what view they took of the affair, for I knew what serious results might come out of that countermanding order. So now you know why I was delayed."

After hearing that story I am afraid I said a lot of things; my blood fairly boiled for a while. All I said was quite in order and more than justified; for those boneheads to calmly and mercilessly condemn us was shocking. After I had let off steam my feelings felt eased. Then I talked the matter over with George and the result of that calm talk was that I took the responsibility of re-engaging Colson and stated that I was the only one who could tell him when he was finished.

I spent the rest of the evening writing out my report to the Sydney crowd; it was hot from the word go. In fact, I had to recast it once. I made inquiries when I returned to Sydney of how the meeting received this report. They had never heard a word of it.

My agreement with the company was that all correspondence between the expedition and headquarters had to go through me. In my report I did not forget to remind them of this agreement and said that all the matters referred to in the letters Fred had brought out were matters I knew nothing of. If they had kept to the agreement we would have got on much faster. I told them that in future, unless any information contained my signature, they were not to take any notice of it, and to get it right out of their heads that a plane could look after the big truck. The boot was very much on the other foot; in fact it would take both trucks to look after the plane. I told them that wherever they were getting their information from it was all wrong and not to pay any more heed to it. I told them that their action in cancelling Colson's engagement might have resulted in the whole party perishing; also that the hot weather was coming on us, and that their interfering had cost us two weeks of good weather we could ill afford. I told them it had cost £100 and also I had re-engaged Colson without loss of time, and the sooner they recognized that I was leader and the manager the better.

I also drafted a lengthy lettergram and gave Fred instructions to hand it to the postmaster and sent a covering letter asking him not to give any mail concerning the expedition to anyone but Fred Colson: I was the manager and was the only one who could give orders or information.

I also received a note from Coote saying that another plane would land at Ilbilba in three or four days. Had we got through to our position we would have been 250 miles from Ilbilba when the plane arrived. I think that would have meant disaster, for as it turned out it was my smoke-signal that guided them to the drome with only a twenty minutes' supply of petrol left. They had seen my smoke-signal twenty miles away. I did not answer this note, just sent a word by Fred that we would wait and to hurry up as I was sick of this messing about.

When we camped at this place we were just about in Mickey's country and I knew from his manner that he was going to break as soon as he got close enough, so when we were making camp I told him not to run away as we would be passing his place tomorrow and I would let him go home and give him lots of things.

He said, "All right, boss, I been worry I fright my leele children, I been on walkabout too long time." So each one gave him something. He cornered-off George and said, "I tink him Archie smokem all baccy, you got some." Of course George gave him a handout, for Mickey and he were great cobbers. Then he told Phil the same tale and got a handout from him. It was a bit of a joke with us, but I thought he might start off in the night so gave him one stick of tobacco. He said, "Two pfeller, one for my lubra," meaning his wife; so I said, "I will give you all this in the morning," showing him about two pounds of very strong native tobacco.

When Fred arrived of course I forgot all about Mick but he was there in the morning so we gave him a bit of something from every case, including a couple of pounds of butter. I told him to give this to Archie Giles. He had quite a load and I mentioned to him to leave it at the house at the homestead and come back with a camel. Seven miles over hills was too much with such a load. I also gave him a note to Archie, thanking him for the loan of Mickey.

There were dozens of four-gallon drums at Ilbilba left by the Mackay party so we loaded about a dozen of these and three forty-gallon drums to drop at Archie's homestead. These drums were better than fivers to Archie for they enabled him to put water dumps out for his men when they were working far from water, the camels taking the water out. This allowed him to work his stock horses longer in the dry belts.

Later on I received through Mr Allchurch all the nice things Archie said in his way of thanking me.

It was funny when Mickey was about to go. He wanted to sit in the back of the truck. Fred told him he could sit in the front with him but he said, "Him tuckerbag might jumpem off." This was Mickey's greatest handout in his life, and he was afraid that he would lose some of it. George in saying goodbye to him said, "Plenty big feller rain come soon then plenty water and grass."

Mickey said, "Yes, me tellem Archie good feller George. How long him come?"

George said, "Two more feller moons, then thunderstorms."

"Yes me tellem Archie good feller George him bring rain alonga cattle."

Later on Fred told us that Mickey talked about this big feller rain all the way home. He was anxious to tell Archie that good feller George sendem rain alonga cattle by two feller moon. It was only about two months later when the country was flooded by several thunderstorms. George will always be remembered at Archie's station as the white feller who sent the rain.

Once more we turned the big truck round and so that there was no mistake I wrote in Freddy's note-book the dates when we would expect him. If anything happened and he could not make it he was to send another truck out.

Fred had taken on a contract to deliver some station supplies when he was put off, so he had to carry out his contract. This meant that he would be three weeks behind us, but as we were going to fool about with planes again the break did not worry me. Again the trucks parted, each going in opposite directions.

23

We Wait for the Second Plane

WE TOOK our time going back and had a look at the country. It was a silent crowd that landed at the bough shed, each going through his part of the work. The waiting was getting on our nerves. I did not make it easier because I was making vows, both aloud and silent, that once that plane came I was going to call the tune. It would have to run according to our plans. I was determined to end the outside control.

George said, "Come up to the soak and have a good sluice and stop worrying about the damn plane."

Three of us, George, Phil and I, had a clean-up and also washed out a change of clothing. It was after sundown when we got back. Harry had been cooking spuds, and as soon as it was dark George mashed them in with some finely chopped onions. It was a bonny feed finishing up with two tins of fruit. And it was grand to eat without being pestered by a thousand angry flies.

Harry was crochety; he was picking on Phil but he got no change from Phil who merely said, "There, that is another sure sign that your liver needs treatment. There are three or four different kinds of pills in the medicine chest. Any one of them is guaranteed to cure liver complaints. Have a shave and wash and you will feel a new man."

After we had cleared away we invited Harry to come over to our quarters and have a yarn. To our surprise he came. It was not long before George led him to talk about his gold reef. Did he remember what sort of country the reef was in? He did remember that the walls were of soft, slaty schist. He said that it was only here and there that the reef outcropped. There were many shifting sandhills and odd patches of mulga scrub. He said that he and Harding walked for five or six miles along it picking up stones at random. When the bag got too heavy they broke the stones up, then quartered them down to about a pound of stone. They did this all along the outcrop. This was the sample they took back and got assayed.

What we were always trying to lead up to was the kind of work he did in the three years he stayed on the goldfields. We never succeeded in getting any information other than that it was any odd job; work was easy to get. On this night he told us some of his American experiences. For four years, he said, he was with a surveyor, then he qualified and got a diploma. I asked him if a foreigner could get a certificate and he said that he couldn't; he had to become a naturalized citizen. Then someone asked him whether that affected him in Australia and had he got a divorce. "Oh no," he said, "I did not use my own name in America. I used my mother's maiden name." This was correct; I had seen his passport and diploma when I had sticky-beaked in his papers the day he left his box open. Although I never had a chance to use the information, had it come to pegging out leases and he still insisted on these being in his name, I would have told the company the truth.

The first and second days of waiting for that fool of a flying machine were not too bad. Phil went to the waterhole and bagged a few pigeons, but mostly we kept in the dense shade of the bough shed. The flies don't like a cool shade and so we were not bothered much. We had read everything readable, so George asked me about the McDouall Peak opal fields. He wanted to know had I done any good there.

Yes, I said, it paid us, but those who came after us, all of them assisted by our camels, they got the benefit of our developmental work. When the Government confiscated our lease, Dick O'Neill and I were putting an open cut through a small hill and getting plenty of indications of opal. Three miners took over and almost from the start they got payable opal, eventually leading them to the largest deposit yet found and yielding £30,000. We were merely the mugs who had done the hard work.

We had meddled a lot with the Hereafter in those pioneering

days and had all of us some pretty close calls. Jim and Dick O'Neill were the real pioneers of Stuart Range and nearly died of thirst on one of their journeys out to it. When I took on a trick bicycle ride of 125 miles in December I got sunstroke, not from the sun's rays but from the heated ground. I was twelve hours without a drink. It took me a month to get over it and after that I did no more daylight riding. Two elderly men perished out there. Their bodies were never found. It was a couple of weeks before we found that they were missing and it was no use looking for them then for the dingoes would have destroyed all traces. A pack of hungry dingoes can eat all the bones of a bullock, excepting the hard leg bones, and the strong jaws of the wild dog would easily crunch the bones of a mere man.

One character came out with a 120-gallon barrel with an axle put right through the centre. He filled the barrel at Anna Creek and one horse dragged it out. The barrel rolled easily enough over the sand and he thought he would get out and have half the water left. Then he intended to let his horse go, but when he arrived he and the horse had used all the water. We had to give them enough water to get away again. The horse had cost him £12, but we gave him £25 worth of water so that he could get to the well seventy-six miles away. We could not afford the water to have him stay with us.

I finished up telling the story of Stuart Range by saying, "Now you chaps have some idea of what opening up a new field is like."

Our minds turned to Lasseter's reef and I told them the story of a prospector who came in with a few fine gold specimens. He was a dying man, all dried up inside. Another fellow looked after him until he died, and the sick man told this chap of the big reef he had found, so long that he could not see either end of it. He told a terrible tale of how his black boy had died of dysentery, how his pack camel had given in at last and he had to shoot it. He had to leave his bag of gold and managed to save only the few specimens. He gave these to the man who cared for him, together with a map of the reef, and then he died.

For thirty years this other man talked about the dead prospector and the reef of gold until he was so bitten by the gold fever that he had come to believe the story and went searching for the reef. After a few weeks of wandering about he too went completely mad and came to another man's place to die where he told the story all over again as if it had happened to him and *he* had found the reef. These poor fellows gather a few specimens from somewhere and keep them for years, gradually building up a yarn that they convince themselves is true.

And I told George and Phil that was what I thought had happened to Lasseter. He had told his yarn so many times that he had come to believe it himself.

That was the end of the fourth day and, with the two days coming back, it was now six days we had lost. I would walk about, going from the shed to the truck, then would mooch over to my heap of dried sticks, pluck a few more green bushes, then back to the shade of the bough sheds. In that dark shade there was some relief from the flies, but not for long. I would go for another aimless walkabout, cursing to myself.

All I could think of was what a waste of time it was with the hot, blistering days coming in on us. So I would rage and boil to myself. To think that after all I had said to the company, they allowed this plane to take priority and control our movements! I took vows on Heaven and the other place that this would be the last arrangement I would make with anything that flew.

George chipped me at last and said, "It's no use you worrying over the plane, it will get here all right when it is ready." I said, "I am not worrying whether the thing will get here or not." I believed that had I been told that it had crashed, I would have thrown my hat in the air. Harry was annoying me also. He had taken one of his religious turns and was singing hymns. I like to hear anybody singing no matter how badly they sing but when one has to listen to a score or more hymns all sung to the one tune, it's a bit trying. If I walked about in the sun the flies would nearly eat me. If I came back into the shed I had to listen to his moaning.

In the afternoon of the sixth day the same five natives with two boys came in. Now I was set. I could go over under the trees where they were sitting and have a yarn with them. They could not understand me and I could not understand them, but that did not prevent us from talking. I would have my say, they would all listen. Then they would have their say and draw many maps, covering each map with a lot of hand-signs. I would just sit and grin and nod my head, not that nodding my head made any difference for I may have been giving them signs that had nothing to do with the sand-maps. They were telling me a great tale but I could not get even an inkling of what it was all about. Natives are not given to talking but these five men seemed very concerned about something. At last they gave in, rubbed out their sand-maps and called the two boys.

The boys were not allowed in the circle while this important conference was taking place. As soon as they came under our shade they were given lollies. It was grand to hear the two kiddies talking, then laughing. They were great cobbers of ours; they could understand far more about us than the men. We all liked to hear them repeating their names—Bubbles and Charlie—they thought this great fun.

George and I made a noughts and crosses' board on the sand and played a couple of games and these two kiddies soon understood. I believed that had they been allowed they would have played that game in a few minutes; but as soon as a board was made for them and they were about to play one of the men uttered one word. Both boys shrank back at once and for a while sat on the other side of the men. Evidently these boys were not old enough to be allowed to draw sand-maps. When I tried to get one of the men to play he just looked surly as if saying it was too childish. They were evidently annoyed with us for not taking notice of what they had been saying.

The next afternoon about four o'clock I heard the drone of the plane. We were all under the trees with the natives. I jumped up, saying that I heard the plane, and ran and got on to my push bike and rode over to my heap of dried sticks and green bushes.

We had heard many false alarms in those waiting days, for the large golden hornet gives a very fair imitation of the drone of a plane. The others only laughed at me when they saw me hurrying over. Before I lit the fire I listened. As none of the others had moved I thought it might be hornets—but no, I heard it again, as though it were banking around.

I set my heap on fire in several places. It soon blazed up; then I applied my green bushes and the dense cloud of smoke went skyward. When I thought I had enough smoke up I opened up the fire and threw an armful of light sticks and grass on it. This flared up ten or fifteen feet. This had the desired effect for it bored a hole right through the cloud of smoke making it into a big smoke-ring and driving it to a great height. I felt very proud of that job but was puzzled that none of the others had shifted from under the trees. Afterwards they told me that they were sympathizing with me for burning up all my kindling that had taken hours to gather. They thought the waiting was driving me dippy; but all the same they were admiring my big smoke-cloud.

After five or six minutes I heard the plane again. This time all hands heard it and before any of them could get over to their heap of sticks the plane was in sight. In a few minutes it was over the drome and made a good landing.

The airmen told me afterwards that they were twenty miles away when they saw my smoke-signal. They were circling over a clear patch low down, examining the ground for a possible landing.

They were not sure whether they were in the right locality and their petrol was on the point of giving out, so they were glad to see the smoke.

When I examined the plane, comparing it with the pretty *Golden Quest*, it looked dirty, scaly and scabby; it had not had a coat of paint for years. I was introduced to Pat Hall, the pilot, who had brought it from Sydney. I immediately asked him where the hell he got such a bottle-oh-looking thing and he said that it was all right, it got you there and it got you back. He added he was glad it was no further as they had only a gallon or two of gas left. He said, "She only does sixty miles per hour all out."

I said, "Well, to think those boneheads in Sydney sent a thing like that," then turning to Hall I said, "You cannot be overloaded with brains to venture out in a thing like that."

Coote said, "Hold hard, we are taking it back to Adelaide to have the engine from the *Golden Quest* put in her and the extra tank. Then she will have a speed of seventy to eighty miles per hour and three hundred miles greater range. They will fix up the fabric and give her a coat of paint. You won't know her then."

The information that they were not staying with us relieved me very much. I then asked why they had not gone to Adelaide instead of coming out here and he said, "I thought I might be of some assistance to you by flying out so you could look at the country you are trying to make for."

"Well," I said, "you can get it right out of your head if you think I am going to fly in that thing from the Ark. If you two are not endowed with any more sense than to risk your lives in it that does not say I have to act the goat too."

So we made it safe and tied it down. Coote was still very groggy on his leg. He was unable to fly the thing so he had brought Hall out. I let him do all the talking that evening. There was a heap of instructions from Sydney telling me what I had to do. In fact, it mostly centred round the plane . . . the pilot would do this, the pilot would do that.

While they were having a feed I said, "Well, now that you have got the thing here, what do you intend to do with it?"

Coote replied in his saucy way, "You have read the Sydney instructions, that's what you have to do."

"Yes, but hold hard. All this looks like instructions from you. I told the boneheads in Sydney not on any account to send another plane, but to let you get some good riding camels and send them out. By these instructions it appears that you have been appealing on my behalf to hurry up and send another plane."

"Well," he said, "I still think you want a plane and still think I can do you a lot of good once I get her out here."

I said, "If ever you come out here in that, bring your own organization to look after you, for I consider you want two trucks and about four men to attend to you. You had a hide telling the Sydney crowd that you could tender the big truck with a thing like that."

But next morning Harry said he would fly in it and I said, "All right, but let's make a chart of the direction you intend to keep and the distance."

So we fixed that up. By this time Phil had given the plane an overhaul and pronounced it OK; so off they went, Hall taking Lasseter on a south-west bearing.

The plane came back one hour forty-five minutes later and Harry said that he had picked a way through the sandhills. He made a sketch and showed it to me. I asked Hall how far did he consider he got and he said, "Between seventy and eighty miles."

Here was a very important point. Lasseter did not say a word to me as to whether he had seen country that he recognized, nor that he had seen his reef.

Next morning when the plane was returning Lasseter handed me a telegram that he wanted to send. I glanced at it and saw at once that it was some encouragement to his wife so OK'd it. This was part of our agreement that no important statement concerning the expedition was to go unless it had my consent. Both pilots were in their places so Lasseter handed his telegram to Coote, then they were off. I felt jolly glad to see the end of Coote, for in my quieter moments I had told him that he was not to move from Alice Springs until he received word from me, and I knew in my heart that he would never receive that word.

At last we were free of the plane. The truck was nearly ready and we got away thirty minutes after the plane had left. We had to use the matting for about half a mile. This was on a soft patch and slightly up-hill. We had the truck loaded to its limit and for the first ten miles I was doubtful whether we had not overloaded it; but Phil did not think so, he said it was the extra heavy going.

The order was slightly changed. We had Harry's sketch to go by and he accompanied me when I went ahead to find a way through. When we turned south we soon found that the country selected from the air was impassable. Harry said it looked good ground from up above; he thought the bushes were saltbush. We had found all saltbush country good going but Harry's saltbush was impassable, so we had to still carry on with the walking ahead.

I think those sixteen miles made the hardest day on the trip; but we were all fresh and bucked into it. We were hugging the end of the Ehrenberg Range and the deep little gutters coming away from the mountains gave us much trouble. We had to fill them up with any old thing we could get and to make matters worse we had to go on an extra three miles to get wood for the camp. This made a late camp, but there was plenty of dry wood. A bit of a cold snap had come and we were glad of a fire. Best of all there were odd clumps of saltbush that meant good going for a start. I pointed this out to Harry and he said, "I am not so sure now, for all that country looked good from the air." There was no yarning that night; lights were out early.

At daybreak we were up and got away by seven o'clock. It was a pleasant day; Harry's sketch worked well. We did not use the matting once but before ten o'clock we had used up both our spare wheels so had quite a long delay. In the next five miles we changed twice. As it was near lunch we fixed up everything again. This was a surprise for we had been on our way for five hours and covered only ten miles. I think we had eighteen punctures that morning. This saltbush had sharp, little, thick sticks sticking out of the old stumps and it was impossible to dodge them. The afternoon was even worse and it was nine o'clock that night before we had mended all our spares. I think we had covered only eighteen miles, but we did not use low gear all day.

The following day was easy on punctures but we had some long runs of matting, two- or three-mile stretches running. As there were only three of us now to look after the four lengths of matting it meant hard work. Sometimes Phil got out and gave a hand. By lunchtime we were all feeling the strain, so George and I had a talk and we decided that we would have to slow down if we were going to last out. Up to now we would roll a length up almost running, then hurry ahead and roll it out. From that time on we worked at a slower pace and did not record our day by mileage, but by the number of hours we were on the road.

I think it was on the fourth day out that we came abreast of a channel that Harry had selected as a possible way for us to get through on our course. We had covered between sixty and seventy miles but had been on a southerly course all the time, so we were very little nearer our objective. I had been paying much attention to our bearings as I wanted to know the distance the plane had covered; so once we started on our south-west course I reckoned up each night how we stood. I got Harry to give me a hand with this because I did not want any arguments about our position when we got 150 miles south of Mount Marjorie.

24

We Lose Count of the Days

FROM HERE I lost count of the days; we had some good patches, and some very heavy going. I think it was on our first day threading through the sandhills that we came to our chief landmark—two razor-backed hills about four hundred feet high. We called them Sutherland Hills, and they were of sufficient interest for us to spend over half a day examining them.

They were a queer mixture containing a considerable amount of very fine golden-coloured mica. George and I agreed that they were not gold-bearing, but might contain wolfram or tin. It looked the kind of country where if mineral did occur, it would be in big quantity.

Here something went wrong with me. I got the shakes, my legs were trembling so that I could hardly stand. Phil was doing something with the truck. I asked him to get my bunk down as I was feeling pretty crook, and he said, "My word but you do look crook." The sweat was pouring down my face. He soon fixed the bunk and opened the medicine chest and gave me a good stiff brandy—about two inches in a pint mug. I stayed in bed until next morning, and whatever it was, it had all passed away; I felt no worse for my trembling bout.

When we pulled up in the shade of some trees next day, George said to Phil, "My word I am feeling crook, I am all of a tremble, you'd better get the brandy out", but Phil laughed and said, "That stunt worked yesterday. You will have to think of something new before you get a handout."

From then on we stopped earlier and cooked proper meals. We all agreed that we were not having sufficient food. We were always too tired at the end of the day to cook, so just had any old thing, mostly out of tins. This matting and digging the truck out was heavy work in addition to the miles of pushing we did. It seems a funny thing to say that three men pushing that truck would make any difference but often when the truck was stopping because the back wheels were slipping, we three would drop off and put our weight behind. If we were quick enough we often got over soft stretches of sand.

On this run we struck some bad patches of cat-head thorns. They fairly made us howl at times. Although this pushing business was very hard work we preferred to do it rather than run the risk of those stabbing red-hot needles. The matting was rough stuff and our hands were worn and very tender—any small prickle was painful. George and I wore out two pairs of boots on this expedition, and the truck-pushing was responsible for half the wear and tear.

The sandhills were running just a point or two north or west, our course was south-west, so we took every break we could find on the south side. One day we made a bad mistake; we had been skirting along dense clumps of mulga when a good-looking channel developed. We took it and went eleven miles before it closed up. We did not find a break on the south side, so while they were turning the truck I went over several sand-ridges about three or four miles across. I found a big open channel so we had to go back those eleven miles to the dense mulga scrub.

We camped there as it had been a fairly hard matting day. When we reckoned up our position that night we had gained only three miles on the air line in a distance of thirty miles. According to Harry's sketch we were being pushed too far north, so next day we ran on to the opening of a big open channel I had seen.

Harry and I then walked for six or seven miles to try out a system of breaks on the sandhills. We found it good going in the truck, but often were compelled to travel three to four miles back east to gain entrance to the next channel. Fortunately we travelled mostly in top gear.

At the end of that day, in spite of our long walks, we had covered forty miles. The air line from our last camp showed fourteen or fifteen miles. Again we camped on the edge of the mulga scrub. We decided that this was the same belt we had struck two days before, so when we were going back we kept on the south side of it and cut off at least fifty miles. The last half mile was really bad. It had low drifting sand that had bluffed us coming out.

Next day we ran up a channel that, according to the sketch, should be an ironstone knob, so we parked the truck under a fine clump of desert oaks, whose shade was as good as any willows.

Harry and I then set out to see if we could locate this landmark. We walked seven or eight miles of easy going and coming back we crossed over a couple of ridges and found the ironstone knob. Harry was very interested in this. So was I, for a start, but after breaking a few stones, I found it was only limestone with an iron coating.

We found a good track going back so we brought the truck up to the knob. The sketch showed two more knobs and next day we found both of them. They were also composed of limestone. I thought it very fair piloting for a man who had not been up in the air before. Harry wanted to walk that day. He said he was interested in the country and wanted to have a look around, so I walked with him. It was while walking up one of these channels that I thought I saw a movement in a clump of tea-tree ahead. I remarked to Harry, "Keep your eye on that clump of tea-tree, I think I saw something move." There was no wind so when we got up to it I had a look around and found fresh tracks of three or four natives. They had been watching us from the cover. The way they had gone was very plain, for they went through a break. I whistled and sang out but they did not show up. We crossed several ridges after that, and when we were making back to the truck I said to Harry, "I believe those natives are following us." Twice I had noticed the flight of the birds behind us. So when we came to a sand-ridge that was covered with a fairly dense growth of shrubs I said to Harry, "We will go over this and when we walk down, tramp a little harder and I will go back and peep through the bushes."

I did this and Harry kept on tramping, and sure enough three natives' heads came up over the next ridge. I was going to let them come over to this ridge, then expose myself and give the sit-down sign, but I did not reckon on their wonderful eyesight. They saw my eye through those thick bushes when they were still a hundred yards away. They turned and ran and I called out and whistled but I did not see them again. Harry's tramping had tricked them as far as hearing was concerned, but not for long as they knew there should have been two sounds. Soon after that we found where two natives had been sitting beside a fire.

Harry had told us about this fire and how he and Hall had seen two natives standing near it. Hall had banked sharply and swooped down low over them, then the wind from the plane covered them with dust and sand. I could see their tracks in the sand where they had run, and I could also see how close the plane had come down over them—their footprints were almost blown out for about fifty yards. Earlier, back at Ilbilba when Hall told me of this incident, I had been annoyed, but now with the evidence before my eyes I was deeply vexed.

I remarked to Harry, "That was a foolish thing to do." He just replied, "Oh, it was only a bit of fun." I said, "By the looks of the tracks the natives did not think it a bit of fun, the prints are easily ten feet apart."

Harry said, "I don't understand you. You went crook about this at Ilbilba and now you are going crook about it again."

"Yes," I said. "If we have a breakdown and have to walk back you'll understand all right and we'll know where we stand then. I think it is essential for us to get in touch with these natives and be on friendly terms with them. You have twice today seen how frightened of us they are."

By this time we were in sight of the truck. They had made camp and had a good feed cooking. The fire was a hundred feet away from the truck for we were getting fairly hot weather now, so were extra careful. I was standing by the fire drinking my pint of tea when George came over. He said, "Harry's lying down. What's the idea of the long walk today?"

I said, "I'm puzzled myself. I went wherever he wanted to go, God only knows why he wanted to climb over a dozen sandhills."

When it became dark we set to on a bonzer feed which George had cooked but Harry did not have any dinner, only a drink of tea. I asked him if he felt ill, but he replied that he was only too tired to eat. We had walked five or six hours in the afternoon. Tonight was the first night that he did not enter anything into his diary and from then on I did not see him writing. He was very quiet; before this if he were quiet, it was a hundred to one that he was sulking and would snap anyone up over nothing; but now he seemed pleased that we were taking some notice of him.

25

We Come to a Halt

BEFORE sunrise next morning we got away and ran the truck about eight miles. Then we had a long spell of matting and pushing from behind, after which we came to a big break in the sandhills. We all examined this. It was very soft sand but the channel looked good so we got to work and bushed and grassed about two hundred yards, then carried the matting ahead and laid it out.

Phil came at our new road at his best speed; he got enough way on, with pushing, to get right through on to the hard ground. We walked back and picked up the matting; this bad patch of sand was over half a mile and we got through without a stop. We had fair going for a mile, only short lengths to mat over, then this wide channel was split by a sandhill, leaving two channels, so I decided to have a walk ahead. Harry came with me. We walked for three hours and came to good, hard limestone, so I sent up a big smoke and the truck came on. When the truck came up we called it a day for it was very hot and there was good shade here.

This country was easy to look at. There had evidently been very fair rain. In fact, ever since we had come in the southern direction the country had looked good. I would think it was six months since rain fell, for nearly everything had seeded. The grass was green at the butts and was mostly of the tussock type. The buck-bush was just green enough to be fair feed but the saltbush looked fine. I have never seen better in my life.

We broke camp early in the morning, but ran only twelve miles before we came to the end of the channel. We had been gaining height for the last three days but in front of us was a flattened-out sandhill that we could see for nearly half a mile; this looked a big walking job. We were all very tired so decided to leave the looking ahead until the following day.

I said, "You can rig up a decent camp for I intend to stay here for a couple of days."

I got up early and went to the top of the long sloping sandhill and saw the strangest sight of tumbled, tangled country. I knew when I looked over that we had come to the big breakaway, and knew that the truck would never look over into that valley. The previous night we had worked out our position and figured that we were somewhere on the border of Western Australia and near enough to being a hundred and fifty miles south of Mt Marjorie. As I looked at the sunken valley I knew this was the end in this direction.

When I got back to the truck Phil said, "I thought today was declared a holiday."

I said, "So it is. I have just been up to see what it looked like from the top." I then told them what I had seen. They all got up while I was preparing breakfast; then the four of us went up, this being the first time we had left the truck unattended. I think George was the first to speak when he saw the extent of this breakaway; he laughed and said, "We can get down there but she will never get out again."

I explained to them what I knew of the breakaway country. As far as I gathered the continent of Australia is a huge flat-topped mountain rising from about three miles from the depths of the sea. Its cliffs are estimated to be nearly two miles of almost straight face. The portion known as the Great Australian Bight slipped away, maybe four hundred miles of the continent slipping into the sea. This caused a big drag inland of landslides, and this breakaway is one of the many folds that occurred. The break that we were looking at was about three miles wide and nearly two hundred feet deep. No doubt Lake Macdonald was part of this gutter. It would be about a hundred miles north-west of us. Lake Amadeus would be nearly twice that distance south-east on the same fault. The break continues to Lake Eyre, Lake Torrens, going out to sea at Eyre's Peninsula in Spencer's Gulf. Stuart Range is part of where the big plain overlapped. On our north coast there is a similar break west of Darwin. I think it would go somewhere between Wyndham and Derby and I am told that the same jumbled country extends through the Timor Island group. So you will see this fault in the earth's crust is somewhere about 2,500 miles in length.

It was a grand sight and we stood there a long time looking at it. Then Harry and I went over about half a mile to a little high knob of flint-stone, only a shallow bed of it lying in the sandstone.

From here we had a commanding view of the country ahead and we sat there and had a good long talk. We estimated we were about 150 miles from Mt Marjorie and were at the position Harry had asked me to go to.

I said, "You can take another shot at the sun from here and make sure," but he would not. He had reckoned up the night before and said we were about 150 to 160 miles in an air line from Ilbilba and the long range of hills we could see were undoubtedly the Petermann Ranges.

I said, "Well, old boy, are you satisfied now that you are bushed? No animal, horse or camel could cross that country." I pointed out that Harding's position could not be south of here as that would bring the total distance from Carnarvon to over 1,300 miles. "In all your stories," I said, "you have never been more than six to seven hundred miles. I have some knowledge of the country south and know that to make a passage through means going back about three hundred miles, then cut across the Winter Glen."

I reckoned that the Petermann Range was no more than eighty miles in front of us but to get there we would have to go back and shift our base to Winter Glen, about five hundred miles to gain no more than a hundred miles.

Lasseter said, "I don't know what to think of your argument about the distance, but my instinct tells me it is over there." He pointed to the eastern end of the Petermann Range.

I said, "But Harry, every expedition that has been out here has gone to Winter Glen. I know of twenty-two. They were all looking for a big reef like you are. The reason why I listened to your tale in Sydney was because it was different to all the tales I had heard before. According to your Sydney story you are six hundred miles too far south now, and yet you want to go still further south. Another thing, you could not cross that country on horse."

He agreed that he had never seen any breakaway country before. I said, "You can understand now why I doubted your story for there was never a horse bred that could cross that country, much less get over twelve hundred miles to Carnarvon. Anyway, why select Carnarvon? There are lots of places closer than that."

He then asked me what I intended to do, and I said, "Had we been six or seven weeks earlier I would not mind giving the north a go because I know there is a mineral belt out there. From here

I would say it is about six hundred miles north-west. At that distance it would bring your story somewhere near what you originally told me, but now that the hot weather is almost on us, I see nothing but to report back to Sydney. You have one slender chance. If we can get this German dogger in you will be able to disappear and save yourself and your family from disgrace."

"Yes," he replied, "I hope we can get that fellow, but some day I will come back and prove that I am no dreamer."

As we were walking back to the truck I told him that I would have a good talk with George and Phil and in the evening we would have a fair and square go at all the facts. Each one would have an uninterrupted say no matter how many times anyone wanted to talk. There was a doubt, and I was quite willing to give him the benefit of it.

26

Lasseter's Last Lookout

WE HAD a decent sort of camp, a clump of desert oak and salt-water willow. We drew the tarpaulin over the top of the smaller trees and this made a dense shade.

The day was Sunday and we were making it a real day of rest. After lunch we stretched out on our bunks and although a word had not yet been spoken each one felt that the job had been finished. It was hard to realize that such efforts had not yielded a single thing. We had come up to this break and looked over the edge as though it were the end of the world. George and I had known for a long time that the story was hopeless but somehow we had the desire to conquer this country; it had got into our blood. We would have liked to have gone right through to the sea.

After we had our nap, George and I went over to the high knob and we christened it Lasseter's Last Lookout. I explained about the Petermann Range; that in thirty-five years there had been twenty-two expeditions looking for this mythical reef; there is a track like a main road. George asked what Harry thought of it now and I explained that he now wanted to go over into Winter Glen, for his instinct told him that his country was over there. George laughed and said, "What, has it come to instincts now?"

"Yes," I replied, "and that's the dizzy limit. I've always found that when a person is governed by presentiments or instincts they are grasping at straws for it is no better than taking notice of a fortune-teller."

George said, "I am sick of talking about these dream millions, let us go down below and have a look at the bottom."

We climbed down under the knob; there were wind caves and many ledges of this fifty-foot undermined face of sandstone. It was composed of coarse water-worn gravel with sedimentary soft iron-stained seams and was very difficult to get a footing on. But when we got down into the bottom we were surprised to see that this had not altered: it was the same as when level with the surface of the plain. The sandhills were unbroken, the channels were clean, still running east and west. There was a fair growth of herbage down here, much greener than up on the surface, and there was an abundance of flowers. The everlasting type predominated in three colours, the white star-like flower that looked like a snow-bush, and a small pink and yellow flower. I gathered some seeds from some good-looking grasses but we could not stay because the flies were nearly eating us and, a peculiar thing, it was hotter than on top—quite contrary to what we had thought it would be. There was a grand sight. South-east of us was Mirage Land giving one of her finest displays: a rugged, mountainous skyline and deep-shaded valleys of tall timbers with flickering patches of water scenes from our movements that made the swift-changing scenery appear as though it was reflected in running water. The Petermann Ranges and other high hills were obliterated by the vertical heat rays. The whole country seemed to be quivering with these, and with the flickering scenes of the Wonderland.

These mirage scenes always fascinate me. They are something I can never grasp. A scene will not stay long enough for one to study it. It will just roll in a blurred mass, then break out into a most wonderful mountain scene, and, before you can look the whole length, it has changed into a scene of trees and water, then a column with dark shades. In its flickering movements one could easily imagine an ocean.

The sun was getting low and casting shadows into this long deep valley. As we were going back to the truck George and I discussed the position. We decided to return to Alice Springs, and, if we could signal the German dogger in, we would give Harry his chance to get away.

That night we held a sort of court. I told Harry that we had decided to return to the Alice but we intended to give him a chance to show us that he had some evidence that his reef really did exist.

I said, "There is to be no argument, each one will have an

uninterrupted say and can speak as often as he likes. If you like, I shall speak first."

They all agreed to that. "Your story in Sydney was that you made the journey with a horse. When I questioned you on that point you said that if you told me all the details I would be as wise as you. Then at the meeting you pleaded that the gaps in your story were because you did not want to disclose the details of your plan at that stage. Up to date, you have not altered that story, in fact only today you talked of your horse. So far you have failed to convince me that you ever made that journey, for not once have you mentioned the big outstanding landmarks of this country. They are its main feature; anyone who had been here before could not help remembering them. Independently of your taking bearings, these mountains should have been your chief landmark, especially as you made the journey twice. I think your behaviour at Mount Marjorie proved you did not know what you were talking about. You took your bearings with the sextant and said that it came out exactly the same as the bearings taken by Harding. You did the figuring in your notebook.

"That night you asked me if I could remember those figures as you had lost them, so you had to go out again next day and take another bearing. Yet, when you took your first shot, you could 'remember' Harding's figures, for you said they came out exactly the same. They were identical with the figures that were placed in the Sydney bank vault though you did not have a copy.

"I have since seen those figures in your notebook. You stated in all your stories that when you and Harding got back to Carnarvon your watches were one hour fifteen minutes out; you did your figuring to make allowance for the discrepancy of that seventy-five minutes. It is a wonder someone in Sydney did not detect that flaw in your story, as a surveyor with a sextant could correct his time every day. I did not think of it until I looked through the sextant on Mount Marjorie. Then your fooling about afterwards, by going out again, was proof that you did not know your sextant.

"You have altered your story about how you were found. First, Harding found you; then you said it was an Afghan looking for sandalwood—he handed you over to Harding. Later on you said that a dogger out after dingo scalps found you. Neither the wood nor the scalps were of any commercial value then, so that yarn is not true. In all your stories you said it was between six or seven hundred miles from Carnarvon; yet, when we got within two hundred miles of your distance, and direction, you decided to go in the opposite direction by a hundred and fifty miles. Now, you want to go another two hundred miles further away, making the total distance 1,250 miles. Worst of all, your only excuse now to go on is that you have a presentiment that it is over in the Winter Glen country; and that is the reason I intend to end this expedition."

I then invited Harry to put up a case why we should go on but he had nothing to say.

I continued, "I will report back to Sydney. If we can engage the dogger, I am willing to let you get away with him, then it will be up to the Sydney crowd to recall you."

Harry seemed pleased that we were not taking him with us.

Morning came and we turned the big truck homewards. This did not bring any great feeling of relief although I had felt for some time that Lasseter was only a dreamer. It was the failure that depressed George and myself; that feeling had to be overcome.

We rearranged the load, and it was at this moment that we discovered that our two cases of lubricating oil were the wrong kind. Phil had asked me to bring the cases to the back of the truck as he wanted a new tin any time now.

These two cases were put on at Alice Springs and had not been moved since. I had forgotten what they were. In fact, this was the first time we discovered that we had an extra case, for we knew that one case was lubricating oil. In my ignorance I asked Phil could he not make it do. He laughed and said, "Just put that in the engine and you will soon have no truck, for it will explode and catch fire."

I admit that I was at fault here but had we been able to go on, it would not have been discovered until the remainder of the oil had cut out. This incident showed how necessary it was to have the auxiliary truck up close. Phil said, "I will try and make some adjustment so that we may get a better mileage." This took a couple of hours.

I put in the time collecting seeds of everything I could. I did this at every opportunity going back. I had a very valuable collection of different kinds of grasses and shrubs, also the everlastings. I listened to a nursery chap and like a big mug gave him all the seeds. He never got any results that I heard of. This still annoys me for I consider that I had a very fine collection. I kept each packet separate, giving the date, the place and distance on each packet, also a description of what the plant looked like. I also gave the estimated temperatures that these plants lived under.

The journey back was very pleasant for there was practically no matting and, best of all, no long walks looking for a way through, but we had a following wind. The heat in the cabin of the truck was almost unbearable. We took every opportunity to run the truck in the shade of trees and let the engine cool down, often changing the water. We were conserving the oil as much as possible. Coming out we were getting only sixty miles to the gallon of oil, but going back we got over twice that distance. Of course, we were in the higher gears and even had short spells of top gear. We could often get up to fifteen miles per hour. George and Harry made a place in the back of the truck to ride, but any benefit they got was offset by being worried with the millions of flies; they were nearly eaten alive.

Next day they rode inside, preferring to put up with the stifling heat rather than the flies. We were now getting a taste of what the hot weather would be like. The consumption of water went up from one waterbag between four of us to one waterbag each, besides tea, for any old time was billy-boiling time. Hot tea is the only thirst-quencher there is. The weather was so hot that even the fire extinguishers were too hot to hold on our knees. Where we could get good shade we pulled up for about five hours during the hottest part of the day, sometimes drawing the big tarpaulin over the trees and truck. The flies were maddening if the shade was not dense enough. The ants were a very real problem also. They would attack in thousands but they would not come in a dark shade.

The most remarkable thing of all was the plant life. This hot weather did not seem to affect it at all. This is the country where the Erosion Commissioners should live and make a nursery—I have no doubt that these hardy grasses and shrubs would save a lot of our erosion problems. If they had men out here collecting the seeds of suitable plants, then went to the landowners and gave them a ton or so of seed to plant in those paddocks that are drifting sand, shutting them up for about two years, the country would be saved from wasting away. I have even gone to the trouble of writing to governments telling them of the wonderful drought-resisting plants (including *parakeelya*) that exist out here; but beyond sending me an acknowledgment, no one has ever taken any notice of my letters. I think it frightened them for it would have meant that they might have to live out here for a summer or two. . . .

On our route there were two big outcrops of milky-white quartz. We had just glanced at them coming out, so we pulled up now and had a good look at them. They contained no mineral of any description, but I took samples. The assay showed that these deposits would be valuable if they were near cheap transport as it was the very finest material for making porcelain ware.

Even dawdling as we did, we were back to Ilbilba five days ahead of the time at which we had agreed to signal the German dogger. We got quite a shock on approaching Ilbilba, for we could see a large mob of camels and what looked like dozens of people walking about. Our first thought was that it was Michael Terry's party but when we got closer we could see that the people were only Rip Van Winkle, his tribe and three other natives wearing clothes who were with the camels.

When we got a few hundred yards further and within sight of the bough sheds the oil gave out so we all went ahead on foot. We soon learnt what all the crowd meant. Billy Buttons, a native lad from Bob Buck's station, and two other natives had come for the empty petrol drums left there by the Mackay party.

Buck had packed the petrol out before the Mackay party came. They gave him the drums and Buck thought that with all these expeditions about it was time he collected his property.

We had some old oil in a tin so Phil took back some of this. We then matted the truck back to its old position.

Billy Buttons was a fairly intelligent chap. He had thirty camels and had made all the empty drums into loads. He was leaving in the morning. I asked him about Rip Van Winkle and his tribe and he said, "I can't make head nor tail of him, we all think he is mad." I asked him if he knew of a bird here that made a track like an emu and he replied that he had seen the tracks, but had never seen the bird. I then offered him a pound note—he knew the value of money—and a pound of tobacco if he would get me one. He laughed and said, "All the tobacco and money would not tempt me to kill one of those birds." He said that they were the blackfellows' religious birds and that he did not want to die yet. I asked him whether he knew what they were like and he said that a blackman had told him that they were about this high (holding his hands apart about nine inches); they had no wings and were a very dark and shiny bird and they could run faster than a horse. Later on in Alice Springs they told me not to take any notice of Buck's boys for they were very unreliable. All the same, the tracks were there, for I had seen them.

That night Harry and I had a talk with Billy Buttons and offered to hire ten of his camels and a boy but he would not listen. He said, "I have been over a month away now and I know I will cop it when I get back."

I then asked him whether he would come back. I offered him ten shillings a week for each camel and £5 per week for himself, but he said, "No, it is too late to be out with camels now. The very hot weather will be in any day. I am going to have enough trouble getting home as it is."

I then asked if he had seen the young German dogger about anywhere and he said, "No, has he been here?" And when I told him he said, "How long ago?"—I told him, and added that I was going to try and signal him in. I said that he might be in any day and this seemed to disturb Buttons; it was quite obvious that he did not want to meet him.

When they were leaving next morning I gave him a letter and asked him to get it to Alice Springs if he could. I also wrote a note to Buck asking him to get it through as we were out of oil and could not move until Fred Colson came. I sent the letters because I feared that something might have happened to Freddie. I took a long chance, but as far as I know the letter never reached the Alice; it was addressed to Mr Allchurch.

This Billy Buttons was a very flash fellow. When they were ready to start he gave us a fine exhibition of riding a buck-jumping camel. I have often seen them buck a little but this was the finest display I ever saw. How he stopped on I don't know, but when the camel had had enough I could still see Buttons trying to make it buck again.

Rip Van Winkle and his party left with Billy.

27

Lasseter Goes with the Dogger

PHIL made a start taking the engine down; all the big-ends wanted doing up. He was three or four days getting it together again. George and I gave a hand wherever we could.

Rip and his family came back after about three days. I had hoped that we would not see him again, but evidently Buttons' camels were too fast for him to keep up with. We were greatly amused by the playful antics of the belle of his tribe. She would walk about showing herself off; then she would put on an old greasy cloth and parade about; in spite of all her dirt she looked very nice. Her figure was as near to perfect as it is possible to get. Her feet attracted me, they had such pronounced arches, but gee! she was dirty. I stood alongside her while Phil took a photograph of us and I can tell you she fairly hummed. I wanted others to stand with her while I took a snapshot but they were not having any of it, for I am sure there were a thousand flies on her. We all thought she would look lovely if she had a hot bath with a cake or two of soap but none of them would take the job on of bathing her.

We had some butter that had got dirty so we gave it to her showing her how to put it on bread, but she did not eat it, she rubbed it all over herself. It made her look cleaner. I think if she had been bathed for an hour or two she would not have been any darker than many of our well-tanned surf girls. She was lighter in colour than my arms or face. Buttons had said she was only fourteen or fifteen years of age but I said, "I understand those two small kids are hers", and he said, "That's nothing, I've seen them with kids at nine years of age."

I said, "They must have been half-castes."

"Yes," he said, "they are always half-castes when you see children with babies."

Phil had got the engine together again, and except for our shooting a few pigeons and having a bath every day, nothing of interest had happened. We had plenty of water at the camp so did not need to go the two miles to the soak. At last the day arrived when I was to send up my signal to the German dogger. I invited Lasseter to accompany me but he gave me a curt refusal. He was paying much attention to Rip and his crowd as he was trying to learn something of their language, thinking it might come in handy later on. I left the camp about 3 p.m. to climb the highest point about five miles away. There were many smoke-signals in the sky that day and I was curious to know if they had any direction, so I left early in order to sit and watch them. I wondered if there were another party coming but concluded that it was just ordinary talk of the hunters coming home to their camps.

The flies were not so bad here about eight hundred feet above our camp and I was able to enjoy another wonderful sunset, then watch the shadow of night blotting out the blue of the sky. At eight o'clock I lit my twenty-minute flare, then went about twenty feet below out of its glare to watch the direction where I thought Johns should be. I stayed until nine o'clock but did not pick up any answering signal so I sent up a rocket. When I got back about midnight the others told me that Rip and his family had cleared out just before handout time, but we did not worry about Rip's movements. The natives were friendly and evidently all those smoke-signals merely meant a fresh movement of the sandhill men.

It was now a matter of waiting for dogger Johns to come to the camp. The young German's story had been that he had run away from a ship and come to the Mission station where he learnt his English. At any time he might be declared a prohibited immigrant. The only work he had done was that of hunting dingoes for their scalps which were worth 12s. 6d. each.

His entry to our camp, in response to my signals, rather antagonized us all. He wore very light boots that were worn through and almost soleless. Nobody heard him come and the first we knew of him was a sharp command to put up our hands while he played with a revolver. It lasted only a second and was his idea of humour but I gave him a piece of my mind that was not music to his ears and had there been any alternative he would not have got the job.

Previously, I had had a long talk with Lasseter about Johns. It had then been very plain that Lasseter did not want to go back to the Alice and we had decided on the dogger as Lasseter thought this was the only choice for him.

After drawing up an agreement and signing Johns on, we set about making up four loads for the camels. He had five camels with him; four pack camels and one for riding. We put everything in crates to make easy handling. There were two with water and two with tucker. It took about two days to prepare the loads. I also mended up my old boots for Johns to take and did a bit of cobbling on his old pair.

By the following evening everything was ready. I had signed the German on for two months, and told him that if he wanted an extension he would have to go to Alice Springs, where instructions would be waiting for him. I also told him from whom to draw his money. My final instructions to Johns were never to take any chances. When he was three days out from known water he was to return and spell his camels two clear days before setting out again.

After we had fixed up the loads, and put them in position, Harry started sorting out his personal effects. This was the chance I had been waiting for. As soon as he opened his tin trunk I went over and took Rear-Admiral Evans's book on navigation. Harry made a grab but I had bided my time and got it. He protested that he could not do without it and made a fine fuss and tried to bluff me by saying that he would not go without it. I told him he could please himself about that for I knew that he did not dare go into Alice Springs.

I said, "You heard me give my word to Captain Houston that I would return this book to Rear-Admiral Evans in person and I intend to carry that out to the letter."

He had already packed his sextant away, as he did not intend to take it, but he was dead nuts on taking this book. For about half an hour he pleaded—I used the book that night for a pillow.

Next morning a hitch occurred. Johns' two mission boys refused to go with him, Ralph saying that four men with five camels would mean two men walking all the time, and the natives knew who those two would be. So these two started off for the Mission a hundred and fifty miles away without food or water. I did not realize that they were going till Johns informed me they had gone. This made a lot of difference, so we decided to drop one load of foodstuffs and take four nine-gallon drums of water. This would enable them to go five days out from the known water.

I questioned Johns as to whether he was a good enough bushman to manage without his boys and he said, "I know the country south from here pretty well and when I get a hundred miles away I will get a boy, for I know almost every one of them over there."

I then asked Harry what he thought about it, and he replied that he did not like it but would have to chance it.

I said, "Most of these chaps who are good bushmen are useless without their native boys." I told Harry all about the country he was going into and not to take any chances, as these light camels would require a drink every three or four days. In four weeks' time they would not safely last three days. He again demanded the book. He considered that he had been "tricked", because, had he known that this book was not to go, he would have brought his own. I did not reply to this because we had not known that Captain Houston was bringing the book.

"I reckon I have more right to the book than you have," he complained.

"But, Harry, as you are not taking your sextant, what's the good of the book without it?"

He retorted, "You don't understand anything about navigation, I hold a Diploma of Survey." Unlocking his little dispatch case, he again showed his diploma, saying, "The name is nothing, my passport is in the same name." At the same time he showed me the passport too, and also a discharge from the American Army. He had often told us that he had fought in the Great War with the American Engineers but from what he said I thought there was something fishy about this American Army yarn; he did not have a credible story about anything in all his reminiscences.

I said, "You have too much lumber now without this heavy book." Just here I want the reader to make a mental note of these facts: he had this strong black dispatch case; inside it there was another smaller case; and these cases were nearly full of papers. There was not room for his big fat diary so it was packed in one of the cases. All the writing papers he took were the same kind, but, when his diary leaves were found, they were on a different paper.

I packed all his personal effects on his riding camel; then I gave him my own prismatic compass, since the one he had was very unreliable. I also gave him my company's watch and revolver so everything was all set. I shook hands with him and wished him good luck.

He said, "I will be able to prove to you later on that I was not telling you any dream yarn."

I said, "If you can find a reef over in that country you are a better man than I think you are for of the twenty-two expeditions

I know of every one was equipped with the best prospectors in the land and they found nothing."

We said goodbye and he added, "If I only had that book I would not care, I feel lost without it."

They moved off, Johns leading the string of camels with Harry walking wide near the last camel. We watched them out of sight; then I said, "That's the end of my millions," for had the ten-mile reef been found my cut would have been worth about seventeen million pounds.

I did not envy Harry his trip, for I don't know of anything more gruelling than to be on a forced march with under-watered camels.

28

We Handle the God

MORNING came and the same five natives and two boys turned up. I was surprised at this for I expected them to bring their women with them this time. They made the boys sit down four hundred yards away, then they came within a hundred yards of us and started making signs. I went over to the sandhill men and drew sand-maps. I soon gave up and started playing with the two boys.

I made a gadget that I used to play with when I was a boy, a round piece of wood about one and a half inches in diameter and about six inches long sharpened at both ends, then a hitting stick. I showed them how to tap the sharpened end of the little round stick; this made it fly into the air rising up about three feet. While it was still in the air, I hit the Jack towards two sticks, the goals. After a while these kiddies mastered and understood the trick and they played several games. Their laughter was worth listening to but Charlie got too good so I altered the Jack so that it would not rise so quickly. Of course Charlie missed it. I laughed at him and called him a "Weywei". The father of the boy then stood up and, pointing to Charlie kept on saying "Ouywel", then would point over the range and say "Weywei". I had at last found a word that we both understood—"Weywei" meant female and "Ouywei", male.

The leader of the band came over to my shade, flattened out the sand and started drawing signs, giving me the male and female signs in the sand, then pointing towards the range. After many hand-signs I understood he was telling me of Charlie's mother. Then he pointed to Bubbles, made his sign, then tapping himself on the chest he put his sign down. Alongside it he drew the female sign for his wife, pulling out a grey hair from his beard and laying it on her sign to show she was as old as he was. In the end I made out that he had two girls and one boy. He went through each one of the men in this manner, then ruled the lot out, and started on signs showing how many members there were in his band or tribe. I counted up and made out that there were sixty-four members including the very old and young children, and I made a note of this in my pocket-book. I felt very pleased over this for in all my wanderings I had never understood what those curved markings meant. I had made a discovery. Sir Colin Mackenzie told me later it was very important and the key to native writing.

After lunch I went over again but lessons were off. They had sent both the boys over under a tree but they were still in sight. These five natives were whispering, making hand-signs that they wanted us, but as soon as I went near they stopped and looked very surly. George said, "Come on, Phil, the natives want to have a yarn."

The natives were whispering to one another in hushed voices but they looked very pleased that we had understood. They formed themselves into a close circle and from a small bag made of emu feathers they first took out a carved stick. The leader of this band made a great fuss, chanting something and then laying the stick on both of his hands and then handing it to me. I could see by their manner that I was being made a grand present. They all stood looking at me, each one saying something and the leader kept repeating his remarks. He pointed to the two boys—then we understood. They wanted me to keep it out of sight of the boys, so I covered it with my handkerchief and put it inside my shirt; then they all yabbered at once and we could see that I had done the right thing for they showed great pleasure at being understood. Then the leader took a stone knife from the bag, placed it on the palm of his right hand, and folded his left arm around his back, each native doing the same. Then he handed the knife to George with quite a long speech. I think I broke a commandment then for I did want that stone knife; the thing was priceless, I am game to bet that George Sutherland is the only white man who has such a knife. It had a carved blade of five inches, and its shape was extraordinary; so fine that its edge was keen as a razor. It had a nice black handle made from the pitch of spinifex grass and up near the hilt was a row of emu feathers. I have searched Adelaide, Melbourne, Sydney and Canberra museums and could not see anything approaching the beauty and high-class workmanship of that knife. It was easily the most outstanding stone implement ever possessed by white man. I later learned that this kind of knife is used in surgical operations, its main function being circumcision.

The blighters did not have anything for Phil. I think Phil was no more disappointed at this than were George and I; but he got a great lecture. They drew him a sand-map, then put their hands on the ground and chanted softly. Phil was certainly receiving a great testimonial, but we could not understand them so could not do the right thing. Phil only had a pair of shorts, a hat and a pair of boots on and he was so sunburnt that he was almost as dark as they.

I think it was because of this that they selected Phil; all this talk and instruction had some bearing on the way Phil dressed himself and it was quite evident that he was to receive something. The leader of the band would take him from one ground sign to the other, then as each was inspected it would be rubbed out. The leader would smooth out a patch of sand in front of Phil and in the plainest of sign language talk with his hands. He wanted Phil to make a sign on the smooth sand so the leader made a sign. Phil took a stick from him and traced the native's sign. They seemed to understand this so the leader made a few more signs, Phil again tracing them. Then the youngest fellow made a signboard circling in several signs. The leader then pointed to Phil's signs, then the young fellow nodded his head, pointed to the sun then to the west. It was quite evident to all of us that something would happen after the sun had set. George and I felt sure now that Phil was to receive something; but there was so much mystery about it all that there was quite a tingle of excitement. George in his own quiet way teased Phil about it saying that it looked as though he would be a married man before the night was out. The natives seemed excited also, as the young fellow went through many actions that seemed to please the other four.

They then went to the shade of the trees, calling to the boys. Neither of these boys had shifted from their position, not so much as one quick turn of the head.

Of course George and I were very pleased. I wanted to swap him his present but the canny Scot stuck to his knife.

We gave them some light refreshments as it was too early for lunch. The men received biscuits and the boys biscuits and lollies. We took the waterbag over and a tin of fruit salts and made a pint of fizz for each one. We could see that they liked this and by their manner were asking for more but we thought it might upset them. We stayed over with them having a yarn about everything but I am sure neither side understood a word.

George was grand cobbers with Bubbles, who would hang around him and attempt to play. When Bubbles first came in he had a few sores. Of course there were dozens of flies on them so George put some ointment on them. This kept the flies away and they soon healed. We thought this might have been the cause of his getting the knife. Quite probably, his kind action had initiated him into the Grand Order of Circumcision. Later on at the Alice I asked a bright native what was the meaning of my stick. He grinned all over his face and said, "My word, good feller that one, you been takem—one woman all a time same as young feller takem black gin of a wife, him good talk stick that one." There was much fun made out of this for I was a married man and didn't know it. I still have the talk stick and when I go back there someday I may use it as a passport showing that I had been honoured by that tribe. I forgot to ask George if he would take his knife and I hardly think he would for fear he would be asked to operate with it.

At the Ilbilba camp later that day when the sun had set we got our fire going and put our tea on to cook. The five men came over to our fire looking quite excited and whispering and patting the ground with their hands. This was the first time they had come over to our fire. We were all paying strict attention to them for it was very plain that they were trying to convey something to us. They did much looking and pointing towards the waterhole. Then this bright, smart-looking young fellow who had been walking up and down, went through his antics. He and their leader were trying to convey something to Phil. George and I were not in on it. When they had done much hand-signing the leader sat down with the others. It was easy to see this young fellow wanted Phil to go with him up towards the waterhole, so after we had a talk about it Phil agreed to accompany him. We could not see any danger in it as he was unarmed but Phil strapped his revolver on. I told him to take his rifle also. Then Phil and the native disappeared into the darkness.

George said, "I think I had better get my revolver." But these men were not paying any attention to us. They were sitting in a row chanting a rather mournful thing and it was quite plain that all of them were excited.

It had become quite dark by this. The two boys had made a little fire; we could see them sitting by it. I had not taken the time but after about fifty minutes we saw the four of them looking up towards the waterhole. Their marvellous eyes had seen the others coming; in a few minutes they were within our firelight. We had purposely not turned on the spotlight, although it was all ready. Phil had run the lead out before he went. We could see that the young native was carrying a long board nearly twelve feet in length and about eight inches wide. It was about an inch thick running to a flat, tapered point on each end, and with lots of carving on it. He placed it on the ground just on the edge of our fire reflection, the other four going over and sitting on their haunches. They commenced chanting, the five of them sitting on one side of the stick each placing the palms of their hands on it—then all palms were turned up as the chanting continued. They were very excited and there was a tremor in their voices. Then the leader looked up to us three and conveyed that he wanted us to sit on the other side, showing where each one was to sit. We three knelt down in our places, then they took our hands, placing them on the stick. They then started off on another chant, much louder this time; the others had been conducted in very hushed voices. We knelt there watching them but it was quite evident that we were not doing the right thing. The leader touched my hand and kept on repeating something. At last we understood that they wanted us to repeat the words after them so now we were set. It sounded something like this "Miney minaver miney he miney me miney he"—there were about five words but I did not get them. After a try we got the hang of it and did it quite well after about fifteen minutes. Then they stopped, and pushing George's and my hands away, they lifted the stick and gave it to Phil and at once backed away into the darkness. Phil stood there and said, "Well, what do I do next?" We were very much impressed and for a while did not know quite what to make of it, so Phil dropped the stick and said, "I'm jolly hungry," so we set about getting our feed ready.

We placed the natives' tucker in two half-kerosene tins and sang out to them but here was something else; the natives did not have a fire, so we turned the spotlight on their quarters—not a sign of them. This was extraordinary.

When Rip Van Winkle cleared out before tucker time we knew there was another tribe handy, but these sandhill men with their tricks puzzled us. Of course the whole of the evening was spent discussing the ceremony that we had taken part in. We tried to get the latter part of the chant but none of us could remember it. The more we talked about it the more we were convinced that something no good happened by taking that stick.

We both questioned Phil closely about where he had got it, and he said that he went up the creek to where it junctioned with the other creek, then they took the left-hand creek (the right-hand went to the waterhole) and about a mile up that the native conveyed to him a message to sit down while he slipped away into the darkness. In case there was something doing Phil went over on top of a little mound and knelt down in some bushes but the native was not long away. Phil heard him coming down the creek so met him. Phil could see he was carrying something but could not make it out. They set off, the native leading the way (I had warned Phil before not to walk in front at any time).

Phil said, "My word that fellow can walk, it took me all my time to keep up with him. He must have been travelling at six miles an hour. Once I bumped the stick and he must have jumped six feet. He kept on whispering something to me. I was jolly glad when I saw the fire again and could not make it out why you did not have the spotlight going."

Before we turned in we discussed whether we should all sleep in the shed. It was decided to separate a little. I went over near the truck although I had no revolver as I had given mine to Harry. But when daylight broke everything was the same.

The stick was still lying where Phil had put it down. In daylight we examined it closely. At first we thought it was an imitation of a diamond snake, the pattern seemed the same; but on examining it each diamond or square was different. From the little knowledge that I had learned from sand-maps, the carvings were very similar to the drawings showing the number of people in the natives' band of five. Now I have no doubt that the stick contained the history of the tribe it belonged to. When I say history I mean the record of their families by blood. There is another aspect of this stick that is worth recording. The wood was tough and light, at the time I thought it was bean-tree wood. These people who have only a stone implement to take a strip of wood off a fairly big log beat me. How they did it I don't know. Judging by the wood rings of each period of growth I would think the tree must have been twenty to twenty-four inches in diameter. Admitting that it was a soft piece of wood to work, imagine the painstaking job it must have been to get that thin strip off. Phil said he would like to take it but the problem was how to stow it safely away. He collected some packing-shavings out of our cases and cut bagging to bind it together.

While we were getting this ready we heard Freddy's truck coming in right up to time. In twenty minutes he arrived. Of course, we three were delighted. We showed him the stick, he knew nothing about it. When Phil told him he was taking it he said, "I might be able to tell you something about it in the Alice." It seems he knew a native there who had come out from this way.

He had good news for me; he had met Ralph and his mate and had given them a waterbag and a gallon tin with as much tucker as they would require to get home. They had discarded their clothes with the exception of their boots and hats. He had met them fifty miles away; that would be fifty miles in less than twenty-four hours. I was very pleased about this for their plight was troubling me.

Fred had camped with Lasseter and Johns that night; they had made thirty miles for the day. I had been certain Fred would be up to time and had wanted Harry to wait a day or two, for there was sure to be a mail, but Harry was eager to push off.

Naturally there was a quiet half-hour while we read our mail. There was only one letter from headquarters; it was full of silly bunk to the effect that I would have to use the plane when it was ready as the flight by Hall had demonstrated how valuable it would be. But that was over so I did not have to reply to it. In another letter I got a shock: Lasseter had wired to the company that he had seen his country and relocated his reef. That was on the day he flew out with Hall. This was the first I knew about it. I read that part aloud to the others and, like myself, they were dumbfounded. Freddy was surprised that we did not know it, for it was common property in the Alice. Now I could understand why Lasseter would not wait for the mail. The little religious blighter had not told me a word; he had tricked me with the wire to his family. I OK'd it but when he handed it to the airmen he could have palmed another wire.

One would have thought that such an important statement from Lasseter would have brought more inquiries, especially with a flying machine on the job, that had, at least, got out there and back.

They could have got a report from the two practical miners, Sutherland and myself, but they did not ask for that; they kept the information for themselves. I did try to find out if there had been any dealings in shares but could not get the slightest information.

Now I had the evidence I wanted, for here appeared to be a clear case of fraud. You can understand now why I had been so particular about anything that went over the air. I suspected that a stunt might be worked, but I was determined that I was not going to be a scapegoat. This telegram was clear proof that Lasseter intended to work a stunt on his own.

As everything was ready for a start we had some dinner and some flies. Phil had stowed his stick under the truck along the chassis. Fred had brought out eight gallons of oil, so we were well set up now. We made a final inspection of the food dump. It was as safe as we could make it. Everything was well off the ground and covered with two tarpaulins. We left all the foodstuffs here excepting what we wanted to get us back. I estimated that there were one and a half tons of food in that dump, enough to last a couple of men for two years. We nailed up elaborate notices telling white men what the food dump was in aid of and asking that, unless they were hard pushed, they would not touch anything, as a prospecting party was out and depended on these stores for supplies.

So we set off—homeward bound—I did feel some regrets at leaving Ilbilba. It had provided many thrills. Just that windbreak and the bough sheds had made it like home; it had supplied me with food for thought.

29

Back at the Alice

FRED went ahead of us taking his truck over the sandhill. He waited to see how we would get over, and as he had a job to do in the Alice, he pushed on, taking the mail also. We had to make some adjustments to our load and took on bags of sand because the driving wheels were inclined to be a bit jumpy. That day we saw where Lasseter, with the camels, had turned off into a waterhole, in fact, Johns was following Billy Buttons' camel-pad. I learned later they stuck to that camel-pad to the end.

I met Johns in Sydney three or four years later and asked him whether they did go on to Buck's place first, and he evaded my question by saying that he could not remember what they did. This camel-pad from Ilbilba to Winter Glen was like a main road—maybe a long way round, but a safe way.

We had a very fair run that afternoon before we camped at our old camp in the timber belt, but we did not lift our drum of water or the food. We thought it might come in handy for someone else, because this was a very long, dry stretch. I think this was the fifth time we had camped there.

Next day we intended to pull up at Mt Liebig and have another good look around but when we ran off our old road we went only about a mile when we struck more big-end trouble. Phil had caught the trouble early and it took only a couple of hours to fix.

Then we put in the rest of the afternoon climbing the wonderful mountain, the view from which I have already described.

On the following day we got only half a mile back on the road when more big-end trouble developed so Phil took the engine down to find out what was causing this. The oil pump was in perfect order, so he made an examination of the oil canals and found them blocked up. He tried to blow them clear with one foot pump—for two days he tried to clear them. It looked like a case of staying there until our time had expired. We had known something was wrong with the engine so had made arrangements with Fred that if we were not in by a certain date he was to come out. The only thing I took from the plane wreck was a staywire about six feet long. Phil cut thin strips off this and bored the channel out. This took days, each of us taking a turn at twisting it. It was a rotten job. We had been caught in a sand-drift and for two days the wind whipped the sand everywhere, but Phil knew his job and got it clear at last. He said it was caused by the breather, which had taken in the sand. If ever he took another job like this he would lead the air-breather through a tank of water so that all solids would be washed out. Quite a simple little thing that would save no end of trouble.

Very little happened from then on, but it was slow. Two of the big-ends had severe knocks. We did not have enough material to fix them. We had four spare big-ends, enough white metal to make eight more, and we had used the lot. So we had to limp in the best way we could. When we could get into top gear or anywhere near it we could avoid the knock. I can tell you it was dreadful to sit behind that engine when it was knocking so badly.

We had a great time at Houston's Soak. We washed our clothes and had two good baths. We needed it, for the job in that sand-hole took six days and had made everything we had dirty. We could not afford to use our water to bathe in because we did not know when the truck would come to a stop.

The run from Houston's Soak was very pleasant. We pulled up opposite Archie Giles' and dropped off a few more big drums. I wrote him a letter thanking him for the loan of his boy Mickey and told him briefly of our failure.

At Hamilton Downs station we stayed for a while and had a yarn with two bullockies who were scooping the dried-up silt out of the tank; they had nearly completed their job. We were a bit of a godsend to them as they were out of tobacco. They had their wives and families with them and we gave the kiddies what tucker we could spare. The women and children looked extra well and showed that this country was fit to rear good, healthy children. I just mention this because southerners are so fond of calling these parts "The End of the World" or the "Dead Heart".

From here to Palm Well it was all good, hard going; we averaged about twenty miles per hour. It was good to feel the old truck moving and none of the dreadful knock. We could have run into Alice Springs easily, only twenty-two miles away, but decided to have a final clean-up and a good bath with unlimited water. This government well is one of the many springs along the Macdonnell Ranges, and somehow, nobody knows how, a date palm came up. There was practically no soil but the conditions evidently suited it. It was between five to six feet in diameter with a score or more of suckers growing out of the base. It had not borne any fruit but was simply loaded with green dates. I learnt later that the fruit matured. They placed some natives out here to watch it and gathered a harvest of over a hundred pounds of ripe, juicy dates. It is a wonder that there are not more date palms through this country; whenever there are any they do well. If ever it is made a national native reserve the Government should put in dams and wells wherever it can. I am sure that at such places practically any variety of fruit or nut tree would do well. Of course you would be giving the natives something for nothing!

We had a bit of a jolly time at this camp. George had a bottle of fifty-year-old brandy. Some friend had presented it to him when leaving Sydney and as he thought we were not likely to want it after the camp we got it out. It was just what we all wanted—we felt fine after it.

Next morning when we had gone a few miles we saw the plane, evidently looking for us, for as soon as the pilot did a few fool things like dipping overhead, he flew back. Tomorrow was the last day of our arrangement with Fred as to the time we should get in.

We pulled up at the telegraph station and gave an account of ourselves to Mr Allchurch. He had a fairly big amount of mail for me, and after reading my home letters I started on the Sydney crowd's stuff. It informed me that I was to hand everything over to Coote as he was carrying on to help Lasseter.

Just as I had finished reading my mail, Coote arrived in a car. He told me that he was in charge of the company's affairs and that as soon as I was ready he would take over, at the same time showing me correspondence from Sydney showing his authority. I replied, "You seem to have been working fast; it is only about twelve days since I sent in word that we were returning. Anyway, that is the position you have been itching for all along, for not once have you complied with the company's agreement."

"In what way?" he inquired.

"By not recognizing me as leader."

When we got back to the Alice there were just as many people to welcome us as when we left. As we limped through the main street, three white men and two natives were all we saw; they did not even give us a second glance. The residents of Alice Springs are used to seeing expeditions come and go. They have had one or more every year for the last thirty-five years. But what they do take interest in is how the expeditions are financed. Each crowd that comes has the same old maps, the same old dead man's tale, the same bags of gold specimens. Our turnout, the best-equipped ever seen in this district, was a little different. We had the original man who had found the dream millions reef; but it cut no ice. It was still the same old tale, only told in a different way.

I was wrong in saying that no one took any notice of us, for when we got into the town proper there was one man who gave us a rousing welcome. It was Fred Colson. He invited us to his home for dinner saying that his wife had a special one; they were expecting us. The natives had reported the evening before that we were camped at Palm Well. Coote was not included in the invitation.

There was no doubt about the dinner being special. It felt grand to have our legs under a table and we had the most wonderful meal ever. The dinner ale was simply grand but we did not exhaust the half of it. As long as I live I shall never forget that wonderful dinner—I did not know how to thank Mrs Colson enough. She was one of those dainty little things who is always clean and natty, the sight of her alone would make a man eat.

We put in the rest of the afternoon reporting ourselves to the Government Resident, Mr Carrington, and visiting the police station where we left Lasseter's big tin trunk.

The old truck's speedometer showed that it had travelled 1,700 miles since leaving the Alice. Our actual travelling distance over new ground was about 1,100 miles, though the Sydney managers and the pretty story-tellers say we had covered six hundred useless miles.

When it got dark Fred said, "You had better get that stick off the truck for I have had a talk with that native and he tells me that even in here it may mean death to any native woman that sees it." So we took it away from under the truck and stowed it in the rafters of his ceiling.

In the morning we started unloading the truck. Coote was there all spick and span, taking notes of everything, playing a lone hand.

When I came to a full forty-gallon drum I said to Freddie, "You can have this as you lost more sweat over it than anybody else." It was the only drum I had marked "Gas". I had marked this drum "Gas" so that I would have a forty-gallon drum of water up my sleeve in case of accidents.

Fred said he thought he had known what it was as he had filled the drum at my request at Houston's Soak. Next day I received a very strongly worded telegram from the Sydney crowd telling me that I would be charged for the drum of gas if Colson kept it. Of course I replied as hotly as I dared, for there was more than petrol involved. So the expenses mounted up; that incident cost the company something over £2.

Phil called me all the mean things he knew. "You had all that water and would not let me have a decent wash." He was referring to the big breakdown where we had been held up for a week.

George grinned and said, "It always gave me a comfortable feeling." He had kept count but did not let me know.

We reloaded the old Thornycroft with the remains of the *Golden Quest*, and what a miserable thing it looked, like a moth without wings. We were now ready to run it on to a railway truck as soon as the station master gave us notice.

As it was a day or two before a train was expected we changed into our city clothes and enjoyed the freedom of civilization; but we were not allowed our freedom for very long. A foreigner came to George and me and showed us some quartz specimens, speckled with fairly coarse gold. He said that he had found them himself and he wanted either financing or to sell right-out. We said that we would see him later and went over to the court house to verify whether he had titles. Not only were his titles right but he had a lot more specimens there. Mr Carrington showed them to us so we dug him up again.

When we started asking for his terms he said, "I only talk in front of Mr Carrington." This made it look all the better, so we went back to the court house and found out his terms. He showed us his position on the map, and I said, "There are some old workings about there."

He answered, "They are a long way west." His titles were all to new country. I told him that I would leave him on the road if he attempted to take me to the old workings for I knew they were useless to us. I knew the miners well who had found and worked them. Again he assured me his finds were to the east. So, after talking to Sydney by lettergram and telling them of the possibilities, we were instructed to make an agreement. We engaged Fred Colson to run us out three hundred miles, but two hundred and fifty of these were on the main road. We reckoned it would take three or four days to look at the place. Of course, the news got around that we were going out again. George and I were in our working clothes ready to leave when Coote came over and said that he was going to accompany us. I told him there was room for only three on the truck and that he had better sit about until we came back. This hurt the feelings of the new manager.

About eleven o'clock we got away, this time going north. It felt grand to be sitting behind something that could move. What a contrast to my last trip over this route, twenty-two years before! When Jim and Dick O'Neill and I had ridden our push bikes from White Cliffs to Darwin all we had to guide us was the cattle route and the Overland Telegraph.

On this trip with the foreigner we went a hundred and fifty miles by the first evening, reaching an old telegraph station built like a fort, with one bachelor policeman stationed there. It had a high stone wall around the place, a memory of the times when natives had once attacked a station and killed the operators. In those days they had to have repeating stations every hundred and fifty miles.

We were on the track early next morning and by lunchtime had covered a hundred miles. Here was the turn-off to the foreigner's mine. I asked his mate, who had come along in a truck, if the turn-off went to the old flat reef and he answered, "Yes." This was just what I didn't want. After about an hour's argument, it panned out that the foreigner meant that his workings were east of the old claims by three hundred yards. So I told him that I was not going any further. We bundled his stores off our truck and turned round, leaving them stand there.

We intended to run right through to the Alice but when we had passed Barrow Creek the wet weather took a hand. We had been racing ahead of a big thunderstorm for three hours and as we were going with the wind the heat was stifling. Inside ten minutes after the storm had overtaken us we were drenched to the skin. There were no doors in the cab of the truck and the rain beat through in sheets. Fred knew this country and said there were two very bad creeks, one about six miles ahead, the other about ten miles, so he kept his old truck going at its limit. We fairly flew the first creek but got caught with the second. There were about two feet of dirty yellow water running in its bed. Fred shouted instructions to us while he was taking off his bonnets. Then, tying a big lump of canvas over the engine and uncoupling his exhaust pipe, he ran the truck into the stream; but the far bank was too steep and the truck stopped. We soon had all the loading off and a big rope out. This we fixed to a tree and the three of us pulled it up foot by foot. By the time we were clear of the creek there was about six feet of water flowing and it was rising rapidly.

Fred said, "I will try and make for Sterling Station. Raining at this rate these creeks will overflow their banks."

We had only three miles to go but George and I gave many strong heaves and pushes before we arrived there. It was grand to see rain after so many months of "dry". The same creek passed within a hundred yards of the homestead and when we got there it was just overflowing its banks. Mind you, all this had happened in thirty minutes. Then it settled down to steady rain and continued for most of the night; but we didn't care, we were under cover.

This was my second visit to the Sterling, and both times it had proved a place of refuge. On that push-bike trip my two mates and I had missed a well by following a camel-pad along the telegraph line until we found out that we were in the ninety-mile dry stretch. We had to cut across and pick up the stock route. It was two o'clock when we boiled up and we did not get another drink until midnight. We did come across a shallow well of brackish water and Dick and I had a drink, much to our sorrow. On reaching the station we were very thirsty and the owner got up and filled us with milk and eggs.

Although the Sterling had changed hands the hospitality was still the same. The new owner, Mr Wright, had his wife out there with him. He, like us, had been caught in the rain, somewhere out on the run. They had one child about two years old; he had no words of his mother's tongue, but seemed quite fluent with the many native children belonging to the station. He was a big, strong, healthy kid and we soon cobbered up.

I think it was on our second day that he picked up one of my words, "bonzer", and he kept on saying it. His mother was quite pleased because it was his first attempt at English. Mr Wright was often three or four days away from home but his wife did not seem to mind; she was not afraid to stay there alone. We stayed three days there. We had only 125 miles to get back to the Alice but it took us the best part of six days. One day we only got a few hundred yards; we were badly bogged three times. What a contrast to our western trip! It was three tired-out, weary men who eventually landed back in the Alice. This time we wanted a bath to remove the dried mud.

Word had come through from Hamilton Downs that the tank was full, also that Archie Giles had got his homestead tank full. This was great news; Archie would be able to save his cows after all. Had the old Thorny been caught out there in that storm I think we would have still been there. Its 3 ton 17 cwt would just about sink out of sight. They had over two inches in that storm and before the year was out they had eight more. That country would be safe for two or three years.

So, once more we changed into our city clothes. Fred informed us next morning that the native boy he had spoken about was outside waiting to see the stick, so we got it down and took its covering off. He seemed very concerned and said it was a sandhill natives' god. The carving on the stick was the whole history of the tribe that it belonged to. He asked how long was it since we got it. Of course we told him how it had all come about. He thought that very likely it belonged to Rip Van Winkle's tribe, and we later found out it did. We told him we had had it over three weeks and he said, "Well, it won't matter now because they are all dead by this, but you shouldn't have taken it." This put a different aspect on the taking of the stick. I went up and reported the matter to Mr Allchurch and he was very annoyed with me and gave me a good talking to. He said, "To think that an old hand like you was sucked in by those myall natives like that." He said it appeared from my story that the stick belonged to the outcast tribe, and he was not sure that that tribe would perish, since they would have the choice of going back under the control of the sandhills tribe that stole it or perishing. I am still very much ashamed of that job. I could have kept it dark and said nothing but I tell it because some day the same thing might happen again. There will be the chance of people knowing my story, and they will not let the myalls use them in this kind of underhand work.

I may state here that Lasseter never saw this stick and had nothing to do with it. He had left the day before with Johns, so did not have the god-stick blessing or otherwise as many people believe.

Phil eventually got it away and I believe he took it to England with him. Perhaps some day, when intelligent whites understand those sandhill men, that stick will have to be brought back. I consider there was no lacking of brains on the part of those sandhill men who put that job over on us. I would call that clever.

30

Our Arrival in Sydney

THERE were a fair number of men in the town waiting for the roads to dry up to move out. I met many old friends and we had rather a hectic time for the next three days. Taylor informed us that he was staying on, saying that he wanted to get all the experience he could in Central Australia for he said it was perfect flying country with the rapid advancement of the flying machine. "I think this will be the first part of the world to be developed by air power," said Phil. "Besides, I am getting £500 a year and everything found and I am staying to get all the experience I can and look upon it as the chance of a lifetime."

Phil's last act with the Thornycroft was to run it on to a railway truck with its queer-looking load. There was no need to ask questions as to what sort of a trip we had had. The Thornycroft looked battered and worn and putting the remains of the plane on it made it look worse. One of the railway men, after tying it down and making it secure, stood back and said, "Well, there is the best load of experience I ever saw," then, turning impatiently away said, "Some people seem to get all the luck in the world; I envy you chaps."

A couple of hours later we were standing in front of our carriage saying goodbye to old and new friends, with a special handshake for Phil while the guard was singing out "All aboard."

So we started on our long train journey to Sydney, something over 2,200 miles. At different times we inspected the old truck to see how it was riding. We found six stowaways billeted very comfortably in the cab of the truck and the cockpit of the aeroplane. They were lads who had braved the north thinking they would get a job; they were allowed to stay until Quorn, the end of the Federal railway.

We were met in Adelaide by a Thornycroft representative who drove us up to Parfield Aerodrome where we delivered the remains of the *Golden Quest*.

Two or three days later we arrived in Sydney, unannounced, and very much unwanted.

Next day I met the directors and gave them a verbal report of the whole trip, sparing nothing nor anybody. I just told them exactly how things stood and of Lasseter's total failure. There was not one redeeming feature in the whole thing. I pointed out how he attempted to do a bit of stunting by sending that telegram. I had seen the copy that Lasseter gave the airmen. It was on the table where all could see it, and I picked it up and said, "You don't want any further evidence than that." But after going through every mistake on Lasseter's part, none of them said a word, not even thank you.

When I sat down the chairman said, "Well, we have heard Blakeley's report, that's all we came here for." Then the meeting broke up. I was amazed at the silence of everyone. While I was waiting for the lift two of them passed me. One put his fist in front of my face and said, "You and Colson won't ever get that reef, we know all about you two, you think now you have got everybody out of the road and Lasseter discredited, you two will go out and find the reef, but we are sticking to Lasseter and we hope when you try and double-cross us that you will perish of thirst." They reckoned I was only playing my cards so that Colson and I could go out later and reap the benefit of Lasseter's knowledge—so I left in disgust.

I wrote my report and gave it to the secretary who said that it would be read at a general meeting in a few days' time. So at last I was free.

I still had one job to do, that was to return the book. I rang the flagship HMAS *Australia* and asked for Rear-Admiral Evans and what I should do with the book. When he heard I had it safe and sound he said, "That is the best news I have had for many a long day."

I asked him where he wanted me to leave it and he said not to let the book out of my hands until he took it from me. So it was arranged that I could call at the Man-o'-War Steps at 10.30 the next morning.

I arrived there just on the minute to the half hour and looked around expecting to see the Admiral. I then told the gatekeeper that I had an appointment with Rear-Admiral Evans. The chap surprised me—instead of answering me he went to the doors leading on to the wharf and in a very loud voice called out, "Mr Blakeley" three times. I had not told him who I was. Then an officer came and saluted me—I was in a quandary, should I return the salute? Anyhow, I raised my hand in some attempt at a salute for I think that was my very first time—what the chap thought of my attempt I never knew. He turned on his heels in real navy fashion and walked before me to the Admiral's barge. There were two officers there and they gave me a very smart salute. I returned it in a fashion but I felt very awkward as to whether I was right or wrong.

The barge was in very choppy water and was bouncing up and down about three or four feet. The first officer offered to take my parcel but I shook my head, and, watching my chance, jumped aboard. I did it beautifully, my luck was in. I had had a fair amount of experience in handling boats in rough water, so I was not altogether a new-chum. I stood for a moment wondering what next, but as the boat was rocking I thought I had better sit down, expecting that something would be said to me, but nothing happened. The barge moved quickly away. As we were passing HMAS *Canberra* all eyes were turned on the barge, I thought they seemed to be taking a lot of interest in it but never tumbled that I was taking part in a grand show. We soon pulled up at a punt alongside the flagship. I think all the bunting and flags on the ship were out, and I now felt really nervous, for there, standing in two rows on the punt, was another lot of officers. This tricked me, I expected them to do some saluting, but nothing happened. Of course I was standing all ready to return the salute. I looked at the officer in charge of the barge, he only gave me the slightest hand wave, meaning that I was to step on the punt. When I did, up went a dozen hands—I returned the salute and again felt very bothered as to whether or not I was doing the right thing—but I kept going between the two lines.

There was another officer standing at the foot of the gangway and as soon as I came within six feet of him he saluted. I returned it and would have liked to have had a good laugh for now I was properly confused. But the two officers came briskly forward saluting me. By this time I was getting pretty good at returning their salute but the marvellous thing about it was that no one spoke.

These two turned on their heels and conducted me to the foot of the gangway, one giving me the office with his hand to go up. Just as I arrived at the top a thought struck me that I should have removed my hat, so off it came just as I landed on the top (a little girl friend had said to me that morning, "Mind you take off your hat before you put foot aboard a warship") otherwise I would not have known anything about it. I always blessed that girl for putting me wise for after it was over I felt pleased that I did not have to be told. The sight that met me there nearly staggered me. There was a double row of differently dressed officers and they raised their swords in salute, then crossed them to make a steel covering down the lane. If I was supposed to return a salute here I'm afraid I didn't, for I had my eyes on a chap at the end of the row who did not have a cap and was dressed as I imagined the Admiral would be, so without looking anywhere else I walked straight up to him and held out my hand. I was greatly relieved when he shook hands for I didn't have the slightest idea whether hand-shaking was right or wrong. He also spoke, giving me some sort of welcome, but the words he used did not stick in my brain. He then tried to relieve me of my parcel. I said, "Are you Rear-Admiral Evans?"

He grinned and said, "No," so I kept the book and said, "Admiral Evans instructed me to deliver it into his hands only." So this chap took me by the arm and led me along the deck. After passing a lot of other officers we eventually came to a door and he took me to his cabin. I can tell you it was a great relief to get rid of that show. The officer offered me a cigarette and told me that the Admiral would receive me in a few minutes. I soon got over my shyness or whatever it was. Here I could catch my wind and actually started talking but not of the show I had just passed through. I explained to him why I refused to give him the parcel and of course gave a bit of its history. After a few minutes another chap arrived at the cabin door, saluted, and said something, then my friend took me by the arm and conducted me to the Admiral. I found out later that the friend was the Commodore. He introduced me to Rear-Admiral Evans and I felt at home at once. For here was a man without pomp and he gave me a hearty handshake, then turning to the others he introduced me.

I gave each one a real handshake and I am still in the dark as to whether I did the right thing. They were all dressed something like the Admiral, I think all in white. There were no frills or pomp here—I felt very much at ease—then, holding out both his hands the Admiral said, "You have brought my book back." I gave it to him, saying that I had had a job hanging on to it for every officer that I met wanted to relieve me of my parcel, but his instructions had been "into his hands only", so I hung on to it. They all gave a hearty laugh over this and while he was unpacking it from its many coverings I nodded towards the Commodore. I had made sure he was the Admiral and nearly gave it to him. I told them that I had asked him if he were the Admiral and this caused another hearty laugh from all. By this time the Admiral had his arms around the book and said, "Never again will you be allowed to leave me." At the same time he thanked me for delivering it.

He said, "I have been worried over it for months, but Captain Houston assured me I would get it back as surely as the sun would rise. When you rang me yesterday I had a good mind to ask you to bring it over at once but instead I gave you this"—waving his hand around the ship. Even then I did not understand that I had had this great reception and that the ship was in full dress for me.

Of course there were refreshments and I had my first cocktails and liked them very much. I gave a brief account of my expedition and talked about my sandhills and the people who lived amongst them and of our trials and troubles in getting the big truck through. The Admiral pointed to the map and said, "What a contrast. That dotted line is the track we took on the Scott expedition." Then he gave me a brief account of their trials and troubles. I could have listened for hours to his account of his trip to the Antarctic, but time fairly flew. Someone announced that it was 12.30—my show was over—so the Admiral asked me to have lunch with him on that day week, and inquired if I would like a mixed party or just plain bucks. I said "Bucks", much to all my lady friends' disgust when they found out. Sir Edgeworth David was also invited by the Admiral to the luncheon party.

The going ashore was fairly easy. When I got beyond the wharf gate I could hardly realize that it was I who had been to such a great turnout. As I stood and looked at the ship I said, "Well, Freddy, that surely was a strange old track you went on that time," but felt as though I had done quite all right. It never dawned on me at the time that I might have done things that I should not have done.

The getting aboard the next time was easy. I presented myself at the gate and was told to go along to the barge that was waiting. So I went aboard and waited for Sir Edgeworth David. It was not long before he came and we were soon aboard the flagship. It was quite an informal affair. I felt quite at home and chatted away with the Admiral, and from the many questions about the trip, I soon got going and gave the Admiral and his eight officers and Sir Edgeworth David a fleeting description of the country, ending up by saying I was sorry to say that there would be no big reef, as

Lasseter was only chasing a dream. Then when it was just about time to go into lunch, the Admiral's wife and her two sisters arrived. They were full of fun and thought it a great joke. They said, "You thought to have a nice little party on your own." I was introduced to them, then we had lunch.

I am afraid I broke some etiquette laws for such lunches for I refused to drink the wine. I asked the Commodore could I have a beer, saying that I knew how much beer I could take but wine had a knack of travelling pretty fast. Then at the end of the luncheon black coffee was served. I shook my head for this was something I knew nothing about, not even what the black stuff was, so sat looking at it. The Commodore was sitting next to me so I asked him if I would be very much out of order if I had another glass of ale. He laughed and said, "That's only black coffee." The others saw there was more fun at our part of the table so my request was repeated. They thought my caution was a great joke, but all the same I had beer. I knew how many beers I could have but a lot of these strange drinks were full of teeth.

Then we all retired to the lounge and continued our chat. I did more listening than talking for the Admiral was extra-special on the Antarctic, more especially as it was first-hand information on the Scott Expedition. I was sorry when it was time to go and as we were parting he said, "What about giving a little lecture some evening on Central Australia?" As I was thinking it over he said he was giving a lecture on a certain date, and what about coming along?

So once again my brain got working and I said, "I will ring you up in a day or two and let you know." That's how I dodged the lecture. I have often been sorry since, but at the time, I thought it would be impossible to speak to an audience. Later on I actually found it just as easy to talk to hundreds as it was to one or two. But for all the worry and trouble I had over that book, I considered it paid me handsomely for I don't know of another in my position who has received a reception on a full-dressed flagship. The funny part about it was that I did not quite understand until my second visit when I was told that the first reception was to the Leader of the Lasseter Expedition.

Captain Houston must have thought as much of me as I thought of him when he gave me a character reference that produced such a turnout. Some of the officers were tickled over the fact that the papers had not heard a word of it but as time has gone on I have realized what a great event it was.

31

My Report Not Discussed by the Sydney Meeting

SOME EIGHT DAYS later I attended a general meeting of the company expecting to hear my report read and discussed. Besides the points I have mentioned before, I had stressed that Lasseter's statement that his and Harding's watches were out one hour and fifteen minutes could scarcely be accepted because Harding was a qualified government surveyor and able to get the correct time any day he liked. This part of Lasseter's original statement had been overlooked by everyone including myself when he was in Sydney. When he took his bearings from Mt Marjorie he had told me that these were exactly the same as those taken by Harding. It was then that I asked him why his and Harding's watches were out, when all they had to do each day was to take a shot at the sun. Had this thought come to me when I examined Lasseter in Sydney the chances are that there would never have been an expedition in search of his fabulous reef.

I also told of the great delay caused by the plane and by the cancelling of the arrangement with Colson without consulting me. The whole party had been exposed to grave risk of perishing. I pointed out that Lasseter was now operating *seven hundred miles* from where he first stated his reef to be. Of those now employed by the company in Central Australia not one had the slightest knowledge of mining.

None of this information was read by the chairman to the meeting. He read something that he said was "extracted" from my report. When he finished there was silence such as might be found in a Sunday school class. Without stopping for questions he went on to other matters. At least a dozen people had heard or seen my full statement but no one raised a voice in protest. The meeting came to an end and I was clean-bowled because no question was asked which would have given me the opportunity to speak. Only two people spoke to me as they left. Those I knew, some of whom I considered friends, openly avoided me. It was one of those remarkable scenes that happen in Australian gold-mining history. A hundred and fifty people were so bitten by the gold-fever that they would not recognize the truth.

The company carried on with what is known as the Second Lasseter Expedition. Lasseter and Johns were out near Winter Glen where Lasseter had marked a spot on the map the evening before he left me at Ilbilba. He would wait at certain points where the plane could keep in touch with him. He would send up smokes to attract the plane once he heard it. As far as I know, no attempt was ever made to keep those appointments, either by the flying machine or by Lasseter.

This second expedition under the leadership of Errol Coote was a costly affair that achieved not a single thing. The fact that Lasseter had given directions as to where the plane was to make contact with him did not seem to matter. Coote dug up a scheme of his own and was quite confident that *he* could locate the reef. Apparently he had a bit of "private information" on which he relied. Instead of trying to locate Lasseter he sent Taylor out to Ayers Rock with petrol and supplies, allowing him fourteen days' start. Then he himself flew to Ayers Rock to join Taylor. He got "lost" and, according to his own account, spent three weeks waiting for Taylor. On the other hand, Taylor spent a like amount of time waiting for Coote. When eventually they did meet Taylor repaired some minor damages which had occurred in landing and Coote flew straight back to Alice Springs. Taylor took the camels back to Bob Buck's place where he dumped most of the petrol and supplies and then he returned to the Alice. Just out to the Rock and back!

At Alice Springs, Coote found that Johns had returned with instructions from Lasseter asking for the Sydney crowd to send the plane to him and saying how urgent this was. But the Sydney crowd had their limits of endurance. Coote had been in charge for two months and all he had accomplished was a costly and useless trip to Ayers Rock. They recalled him to Sydney notwithstanding Lasseter's urgent request. When Taylor landed back in the Alice fourteen days later they instructed him to load up more petrol and supplies and proceed with Johns to Ilbilba by camels. A plane was to follow later with a new pilot and the three of them were to search for Lasseter from Ilbilba. It was an extraordinary way to treat an "urgent" message for help.

I know how easy it is to be wise after the event, but let us go over the position again briefly:

Johns returns with the information that Lasseter claimed to have relocated his reef. Lasseter was then only a few days away from Alice Springs instead of the three hundred miles west of the town, where he had originally claimed to have found his reef.

Coote's plane was at the Alice and Johns was there to guide the pilot straight to Lasseter. That could have been done quickly, especially as they had a dump of petrol at Bob Buck's place, closer than Ilbilba to work from. Had they adopted this plan, they would have caught Lasseter napping for he had stated that he had seen his reef and was returning to peg it. Still, you never know what men smitten with gold-fever will do. It is seldom the obvious or sensible thing. The Sydney crowd took their time—about a month—and then sent out a new-chum pilot who promptly got himself lost. After another month Taylor was sent an urgent message: "Go out and bring Lasseter back to Sydney." (This is recorded in Phil's diary.)

The message was sent three months after Lasseter's urgent appeal for help. Taylor started immediately but got sick. While on the search he sacked Johns and then he handed the search over to Bob Buck. Buck's entry into the search added to the mystery.

Let us now trace Lasseter's and Johns' movements. When they returned to Ilbilba, Lasseter sent Johns into Alice Springs with the mail. Johns arrived about 6th November, 1930; and close to that date his agreement expired. Among the messages he carried was an extraordinary letter to the Chief of Police, requesting that Johns be arrested or held until Lasseter returned. The charge against Johns was for some sort of violent assault.

Lasseter stated that he would be back within two or three weeks. He had found his reef and was going to peg it and bring in samples. This was the gist of the information he sent to the company in Sydney. He also stated that he did not trust Johns, so had sent him to Alice Springs. He said that he had got to within a mile or two of his reef and was going back with two camels and fresh supplies from our dump at Ilbilba. Of course, the Government could take no notice of Lasseter's letter and the company re-engaged Johns for fear that he knew too much and sent him out with Taylor and the camels to join Lasseter. Here had been the golden opportunity to test out the plane. Johns knew the country and could show where they had turned back when Lasseter had stated that he was only a mile or two from his reef. All that was needed was to fly out with Johns and contact Lasseter. The plane could have flown out and back in a day. Lasseter would have been no trouble to find and in that open country there would be many suitable landing places. Johns could have been left in charge of the claims while Lasseter was flown back to register them.

But, as I have said, instead of using the flying machine the company sent Johns and Taylor back with camels. But they could not "find" Lasseter, as the flying machine could not "find" him. It was a wonder the whole blanky lot were not "lost". The end of the Taylor-Johns trip was that Johns got "bushed" and had a hard job getting back. I ask how can a man get "bushed" in country with four or five well-known landmarks scattered about, a man who had been travelling the country getting dingo-scalps for two years? He had well-worn camel-pads that were used more by his own camels than those of anyone else. Johns could have gone blindfolded to where Lasseter was supposed to be. Johns was better equipped than any other bushman for this bit of country. He could speak German and English fluently and had a good working knowledge of most of the tribal languages. There was not a waterhole for hundreds of miles that he had not visited. He was better known to the natives than any other white man. There was scarcely a native with whom he could not find a word or two of understanding. All these natives were more or less under the influence of the Hermannsburg Mission and quite a number could talk a bit of German or English. I feel that Johns must have had trouble dodging the many camel-pads from one waterhole to another and the three well-worn pads leading to the mission station, Winter Glen and Ilbilba.

What reasons could Johns have had not to take Taylor across to Lasseter's camel-pad? Was he afraid to? Was there any agreement that fitted in with the secret telegram?

Finally a hue-and-cry was made about Lasseter being "lost". How the report got about I never found out. The gold company was in the limelight. A plane was chartered to search for him, though the two airmen's knowledge of the bush was nil and the two pilots made a full load for the machine.

The airmen set off full of confidence that they could locate Lasseter in a day. They had never seen the sandhill country before and had a bad attack of aerial whirligigs and were promptly lost. The speculators were in the seventh heaven of delight for they now had their gold-fever victims howling to everyone for assistance.

Big headlines appeared in the newspapers every day. In the anxiety about the lost airmen poor old Harry Lasseter had to stay very quietly "lost" until the limelight was turned in his direction again.

The Government sent out part of the Air Force, four planes, which cost the taxpayer over £3,000. The first two planes left Point Cook on 1st January, 1931. One was flown by Flight-Lieutenant Charles Eaton and the other by Flight-Officer "Jerry" Gerrand. This was not the first job of its kind that Eaton had handled, for it was he who had found the bodies of Keith Anderson and H. S. Hitchcock after their tragic loss in 1929. The planes were equipped with large auxiliary tanks to give them a wide cruising range, and they scoured the country from Alice Springs thoroughly for almost a week but with no results. Then Gerrand's plane hit a concealed stump at Ilbilba and was put out of action. On 7th January two more planes arrived from Point Cook.

At last, Fred Colson and Archie Giles were enlisted by the authorities. Fred was off within an hour, taking with him a good native who could make and read smoke signals. Within three hours he was a hundred miles away, talking and planning with Archie Giles as to how they would comb the country. Inside another hour Archie was following Freddie with a team of smart natives mounted on their fastest camels, any one of which was capable of doing a hundred miles in ten hours.

Fred rattled ahead with his throttle wide-open making row enough for two machineguns. Anyone in that sound-carrying country could have heard him as he roared by on his way to the first waterhole at Dashwood Creek. There he filled his spare drums, took a drink of cold water, and went on non-stop to Ilbilba. Archie arrived ten or twelve hours later at the waterhole. They intended to spread out, keeping in touch with each other by signals. The sharp-eyed natives immediately spotted fresh tracks that were not Colson's. They showed these to Archie and then their searching eyes looked all around. They pointed up the creek and told Archie, "Two feller white man sit down alonga bough shed." Archie went off in the direction indicated and there found the missing airmen.

In some extraordinary manner they had not heard Fred's truck go by, so they declared. Maybe he was making too much noise! They had not made and kept fires burning or they would have been detected days earlier by Flight-Lieutenant Eaton who had passed *right over their heads*. In his report, Flight-Lieutenant Eaton commented on this neglect to keep a fire burning when there was plenty of fuel and matches and stressed the necessity for this being done in such circumstances.

Shortly after Archie found them Flight-Lieutenant Eaton was back again. But consider the comparative costs of the land and air searches. The ground search would not have run to three figures, but the air search cost £3,000. It would have got more expensive as it continued for this country is hard on planes, but with the land party the first expense was the heaviest. The air search depended on persons sought not only being alive but willing and sensible enough to send up signals. The land party, with the aid of natives, could and did find the people. If they had been dead the natives would still have led the way to their remains.

32

Weaknesses in the Story of Lasseter's End

NOW that the flying sensation was over the company could turn its attention to searching for Lasseter. But did they keep Fred Colson and Archie Giles for the search? Not on your life. They were afraid that Colson and Giles might pinch the "reef".

They hired Bob Buck. I don't know how long he took but, in due course, he reported finding a body, burying it, taking a photo of the grave. He sent in a few pages written in lead pencil saying that he found these on the body of Lasseter. I was in the Sydney office when they arrived through the post. There were only three or four pages, telling how Lasseter's camels had bolted. He claimed that he had wandered about with a tribe of natives for days, then dysentery and sandy blight had set in. All but one native deserted him. He described this native as having a big wart or loose lump of skin hanging from his behind, and asked those who found him to be kind to old Warts.

On another page he said that he had pegged his reef and gave a vague description of its position. I queried the handwriting and the paper and suggested that it be submitted to an expert. I thought that those pages came from a very cheap notebook, something like what kiddies buy for a penny. All the paper we took on the expedition was very good, and I was also sure that these pages did not come from Lasseter's diary which was composed of very good quality paper. I am almost certain that the pages were not submitted to an expert, because the gold-stricken people feared that I knew too much. I had seen Lasseter's diary and had had in my possession for three weeks the map that he had marked showing the position of the small lake and the hills called the Three Sisters. Between that lake and the Three Sisters was supposed to be the position of the non-existent reef.

The Sydney crowd were in a high fever lest I gave this imaginary "information" to someone who would get there first. Meeting after meeting was held. They sent for Bob Buck to come to Sydney while they were reconstructing the company on a bigger scale, on the strength of the statement that Lasseter had pegged his reef. No attempt was made to verify the story, though they had the *Golden Quest* at the Alice. A report could have been made in eight or nine days by using the air to Alice, then flying on to the mission station and ordering a truck to be out there to receive the mining expert, with a few light camels sent ahead. The report would have been something definite upon which to form the expanded company. But no, they sent for Bob Buck.

I was having a day out in Sydney some time later and having a drink in a hotel when in walked Buck. I asked him to have a drink, and said, "I know you well enough but can't find a name for you." He said, "Why, I'm Bob Buck from Central Australia." Now everything was clear. He only wanted to talk about cows and bullocks but I said to him, "Look here, old man, you never found Lasseter's body. Harry was too well equipped to go out the way those diary leaves said, and why was it that you did not find the two camels and the loading?"

He said, "God only knows where they are."

I said, "You know as well as I do that a bolting camel would throw its loading within three hundred yards. That's a very thin yarn to say you could not find them."

He then asked me who I was, and I told him. He was greatly upset for he had had many warnings not to talk to me. "You won't crab a fellow's pitch, will you?" he said.

I was pretty disgusted and left him but he followed me out into the street and told me of the hard times he had had, and that he had lost 1,100 head of cows through the drought. "If I can hang on to this game for a few more weeks," he said, "I will be able to restock when it rains." At the time I felt inclined to put the whole show away by going to one of the papers and making a statement, but no matter how gently I handled the affair then, the mother of Lasseter's four children would suffer most. I might have kept quiet for ever if it was not that the Lasseter Reef story refuses to die.

People are still losing money over that pretty tale, so it is time the truth was told.

My first recollection of the fabulous gold reef in Central Australia dates back to 1899. I was in Fremantle, having come from Broken Hill, where I had been attracted by the big money to work on a fire-screen underground in the big mine. The underground workings were on fire and I took on a job at £1 a day. In a few weeks I was full of lead and nearly a wreck. My doctor told me to go to Fremantle where, in that bonny climate, and with my youth, I had a chance of recovering.

While at Fremantle I met two miners from the Western Australian goldfields. They were also at Fremantle for health reasons; they had saltwater boils, frightful looking things with core holes as big and deep as my finger. When they were cured they were going back to sink some Government wells, and expected to make good money. But the main reason they wanted to get on this line of wells was so that they would be there when good rains came. Then they came out with the story of how a surveyor named Harding had found a skeleton way out on the scalded plains. They knew the position near enough. This skeleton had a bag of rich gold specimens alongside it. In fact, the white sun-bleached bones of the dead man's arms were still coiled around his bag of stones. The two miners had seen some of the stones and they said that it was very rich stuff. Harding was of the opinion that this man could not have walked very far but in the course of a year or so the surveyor had found no traces of the reef. He had taken good miners out with him, but they could find no auriferous country.

Harding was a great pal of Sir John Forrest. When Forrest became Premier of Western Australia, Harding was so persistent about this reef that Sir John gave him a well-equipped camel expedition. It was a big party and they were away about seven months but they did not find the reef from which the bag of stones were thought to come. All they found was the skeleton of a horse with an old saddle and a quartpot alongside it. The dead horse was nearly three hundred miles further east, but Harding believed that this skeleton horse was part of the plant that belonged to the dead man of the bag of gold specimens.

My miner friends went out and dug their wells and wrote to me saying that good rain had fallen and they were going after that reef. I replied telling them that I was not a starter, for at the time I thought I would never go down another mine. They stayed out all the winter but, of course, found nothing. Some five or six years later good rain fell again and they went out once more.

They went too far. The waterholes dried up and cut them off, and a couple of years afterwards they were found lying side by side under the shade of a saltwater willow-tree. They didn't have a bag of stones or any diary leaves, but Harding who had the gold-fever badly all those years, left the government service and kept hurrying out to get ahead of other parties until it killed him. When he died that story became a dead man's tale that cost several lives. It is a hundred-to-one probability that the skeleton with the bag of stones was that of some poor old prospector who had come from the main goldfields. It is possible that he too had started telling other fellows dream yarns about these stones coming from far away in the Never-Never and told the tale so often that he came to believe it himself.

Some thirty years after Harding found that skeleton, Lasseter came to light with the story that *he* was the man who had found the reef and that *he* had been found by Harding nearly perishing. Lasseter had his bag of stones, and Harding had been dead twenty-five years when Harry turned up with his story in the Sydney office of the Australian Workers' Union.

Let me give you a straight tip. When you hear a fellow say that he knows where a reef is, "ten miles long, eight to twelve feet wide", call him a liar straight out. According to one writer there have been eighty-two expeditions out after that reef where that dead man's bag of stones were supposed to come from. I used to be able to count up to twenty-two expeditions that I knew of myself. No other part of Central Australia has had such a combing. First-class prospectors have travelled hundreds of thousands of miles and none has reported finding a yard of auriferous country. A dozen capable geologists have examined the sandhills and all their reports are in agreement. It is impossible for this country to bear minerals. In my report I added a rider: the country may contain oil. There are indications that bed-rock would be seven thousand feet below the surface and oil experts say that there must be a large natural basin to hold oil. Well, the basin is there with a surface of hundreds of miles. The flying machine is retarding oil exploration because the experts too often want to prospect for oil from up above the clouds. They come down to make elaborate maps and say there is no oil, but they have never been within ten thousand feet of the surface. In time this will change and land parties will study the cause of the blowholes. As far as I know no one yet has taken a sample of the stuff that comes up under pressure from those whistling, whispering blowholes.

I am writing this story where I have been doing a bit of prospecting fifty miles over the Great Dividing Range from the bridge over the Delegate River. It is Bill Mead's place, and has an elevation of 3,800 feet above sea level. I came to do some mining but a heavy fall of snow, about two feet deep, stopped me from getting below the surface with my pick and shovel, so I started to write this story with pencil and paper. For two months I have done nothing else. I have told here how I believe Lasseter's story began, and I think you might care to know my views about his presumed end, for I suppose that I have thought more about this than any other living man.

Now, the story as told on those few pages brought to Sydney, was that Lasseter's camels took fright at something and ran away from him. Those two camels were loaded to the extreme limit and had been working continuously for about eight months. Both of them would be very tired beasts long before that. According to the accepted story the camels galloped away out of sight with the pack and Lasseter saw no more of them. As I have pointed out, the packs would not stay on long no matter how they were tied. Inside a mile the beasts would be so exhausted that they would slow down to a steady walk. It was merely a matter of following them for maybe less than an hour before Lasseter could have caught them again. It appears to be Lasseter's story that he just stood and watched them run away. The natural thing for any man would have been to follow them, as his life depended on the water and food that was on those camels.

Instead, Lasseter is supposed to have started wandering over the sandhills looking for unknown water and food. He wandered for a day or two, then he came into contact with a tribe of natives. All he had when he met them was a dry waterbag and a revolver.

But to return to the mystery of those camels. The two camels were bred on the Hermannsburg Mission about a hundred and fifty miles away. Even with their packs still on they would have made straight for the mission station, arriving there in about four days. Without the loads they would have made it inside two days. Now the missionaries are most humane people and would have acted at once. They would know that a white man was with the camels and would be in need of assistance. They would get their fastest camels and send a supply out with reliable natives, tracking those two camels and finding out what had happened and where it happened. Lasseter would have been relieved within five days, had the camels turned up without their loading; in nine days if they had come home with their loads. On fast camels, the natives would have been able to track the camels that had bolted, maintaining eight to ten miles an hour. The natives can see stones the size of sugar that would be displaced by the pad of a camel's foot. Camels do not perish because there are too many open waters in this country. A camel can smell water, if it gets to windward, ten miles away. They are as sure as homing pigeons; they always make back to where they were bred. Had those camels run away they would not have stopped until they arrived home at the Hermannsburg Mission. They would never have been killed and eaten by natives for the same reason that natives will not eat horseflesh. *But those camels never came home.*

What happened to Lasseter's property? There was his strong, steel, black dispatch box that would be locked. It contained the great fat diary in which he wrote daily in camp. It contained the seven gold specimens and all his papers. All natives of this country have at some time or other come under the influence of the Hermannsburg Mission. Hard, long drought years drove them into the Mission or else the missionaries made depots out in their land. A few of the natives always stayed with the missionaries, mostly children who, when they were old enough to fend for themselves, would go back to their tribes. In this way there was not a tribe in the Winter Glen district that did not have some member who could speak a little English or German. They knew the value of "reward" and all would understand that if they took information of a white man in distress to the Mission, the reward would be a big one. The Mission is generous to any natives who bring in messages of people in distress, whether these are white or black.

If Lasseter had linked up with a tribe as he described, the elders or leaders of the tribe would hold a council and select two young men to run in with a message. If Lasseter himself was incapable or could not understand that they wanted him to send a message, they would themselves carve a talk-stick. Within six days he would have been relieved. The two soft-footed sleuths sent out by the company spent months in which they did little else but look for that dispatch case because they knew it contained specimens. Of course they did not know the specimens had been bored and loaded and could not be used for another "find".

I myself packed Lasseter's pick, shovel and axe in his load, and in his last, lingering days when he found his reef, he is said to have pegged it with hardwood posts. How did he manage to cut hardwood, dig holes and trenches without his tools? Why were those tools not on the claim when Bob Buck found it? Why were they never found? Why were not that fat diary and his specimens found, together with the dispatch case? The simple reason is, I am convinced, because *they were never lost*.

Why did not Buck look for the lost camels and the loading?

There were good plain camel-tracks. Buck had command of all the natives around there. If he did not know their language he had his own natives with him who did. A few smoke-signals sent up by Buck's natives would have been sufficient to locate the loading or camels within an hour. They might be a hundred miles away but the direction would be definite. Nothing like that is ever lost in this country. I have told how the natives keep the news of your direction written in the sky; they always watch those who enter their tribal countries.

Who supplied Lasseter with the common paper? I am sure that he did not have any such paper with him, for all our paper was good stuff.

The story of Lasseter having dysentery and sandy blight resembles a yarn I once told him. In spite of all his yarns about how he was out for months with that good bushman, Harding, he had no bush knowledge. This statement could be confirmed by the other members of the expedition, George Sutherland and Fred Colson, both good bushmen. Phil Taylor's diary confirms it.

So many incidents occurred that disgusted us, and when he pretended to knowledge he did not have, he only made things worse. Bushcraft is something that takes years to acquire and it can only be learnt by practical experience. Once when we first got into the sandhill country Lasseter kept telling us that he had seen wild turkey dung. This puzzled me and I said, "Next time you see this turkey business you show me." So when he drew my attention to it I saw that it was the dung of a kangaroo rat. I said, "A nice sort of bushman you are. Why don't you look at the track and find out what animal is responsible?" There were many little incidents like that one. Almost everything Lasseter said or did was the act of a raw new-chum.

Now, frankly, I do not believe Bob Buck's story of the finding of Lasseter's body. I believe that the start of this story was in the mail that Johns took in from Ilbilba. Otherwise, how could everything dovetail so perfectly? Johns's story of being "bushed" in country so familiar to him and in which there were two main roads that he well knew, to say nothing of the landmarks, just cannot be accepted. It is a funny thing that Lasseter's body could be so easily found when the larger things, that had camel- and man-tracks to them, could never be found. The death of the man was never investigated by the police. They received Buck's report, and from what I can learn, they did not believe it. I suppose that Lasseter is the only man reported to have died in abnormal circumstances in the Centre whose death was never investigated by the police. I ask: Why was it not investigated? Bob Buck's story is much too thin and, with so much in doubt, I cannot believe it.

One final point. That last diary that came to light and was used so effectively was quite recently written and undoubtedly in Lasseter's handwriting. I should greatly like to know the watermark on the paper and whether experts would be able to trace it to Mormon headquarters.

Now let me finish off the story. The new company had one active businessman who had invested a large sum of money. It was he who suggested that the company ask the Western Australian Government to lend two of its geologists to accompany the third expedition. In due course the party landed on the ground where Lasseter is supposed to have pegged his claims. One look around was enough to tell the geologists that there was no auriferous country within those pegs. In their brief report they stated that none of the country searched by Lasseter was mineral-bearing. They very quickly ended the expedition by describing the whole country as non-auriferous.

Since then almost every year government or private geologists have tendered similar reports. Here is a news item from the *Sydney Morning Herald* of 8th July, 1936:

> Perth, Tuesday: Another Lasseter's Reef theory was exploded today when the Minister for Mines (Mr Munsie) received a telegram from the Government Geologist (Mr N. A. Ellis) stating that no reef existed in the locality reserved by Border Gold Reefs Ltd., Sydney, the country being sandstone.
>
> The company's reservation, which was dated to 30th June, and then extended until Mr Ellis made his report, is now annulled. Mr Ellis returned to Alice Springs today.

That, my good friends, is as true an estimate of Lasseter's fabulous reef as you will ever get. For the last forty years men have been finding directions in pickle-bottles of where this reef is supposed to be—that was long before Lasseter left school. And the two hard-heads who went away for a short time and came back saying they had found leaves from Lasseter's diary in a pickle-bottle planted under an old fireplace, were merely following the good old tradition.

The company had several options over gold-mining claims so these kept them going for a time, but with their depleted banking account the gold-seekers fell away, and some once more became sane citizens. A few even say good-day to me now.

When I set out to write this account I did not intend to hurt anyone's feelings but to stick to the facts. From a mining point of view the famous Lasseter expeditions were a total failure. The same can be said of all the other parties who went looking for the hallucinations of cranks who told their lies so often that their brains were affected and they became the victims of their delusions.

So be warned if ever you have the bad luck to find a dead man's bones with a bag of stones, or an old diary—be very careful. Burn those old bones! Bury the stones! Use a long-handled shovel well greased so that the gold-bugs won't crawl on it, for their bite gives quite sane people the gold-fever that often proves fatal.

From my own point of view the expedition was a great success. We did not find those dream millions, but found something more precious. We found the Sandhill People. This book will justify itself if it brings some response to my plea for the preservation in their natural state of those lovely people of the sandhills.

Index

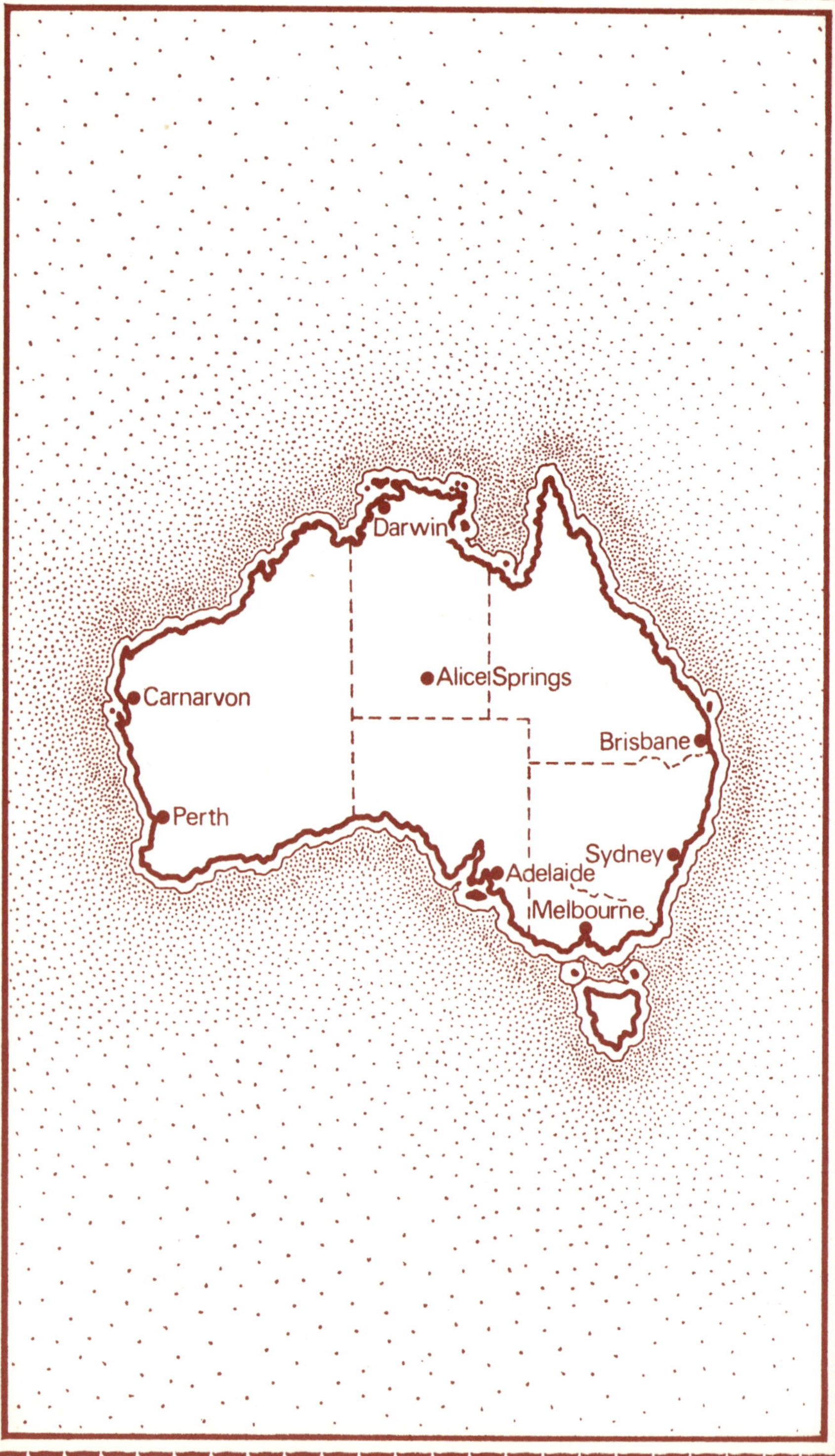
Darwin
Alice Springs
Carnarvon
Brisbane
Perth
Sydney
Adelaide
Melbourne